Contents

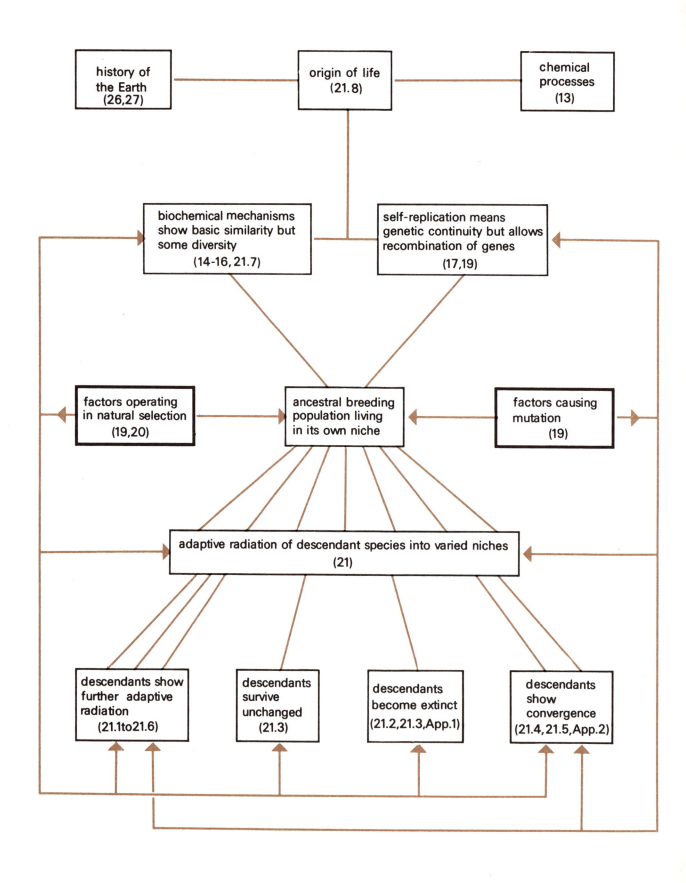

Table A

A List of Scientific Terms, Concepts and Principles used in Unit 21

Taken as pre-requisites			Introduced in this Unit			
1 Assumed from general knowledge	**2** Introduced in a previous Unit	Unit No.	**3** Developed in this Unit or in its set book(s)	Page No.	**4** Developed in a later Unit	Unit No.
General characteristics of: fishes frogs lizards and snakes birds mammals apes poppies columbines snapdragons bees General features of design of aircraft and submarines	mutation theories of evolution natural selection definition of species and genus niche distribution of organisms control of body temperature cerebral cortex chimpanzee behaviour echo location by bats moths caterpillars and other insects crustaceans (water fleas) visible and ultra violet light unicellular organisms biochemical processes in various types of cells autotrophes and heterotrophes cytochrome c control of protein synthesis photosynthesis teleology genetic code	19 19 19 19 20 20 18 18 1 2 19, 20 20 20 2 18 14, 15, 16 20 15 17 16, 20 18 17	evolution as the gradual accumulation of changes (mutations) speciation of Galapagos finches evolutionary history of vertebrates adaptive radiation of vertebrates, including swimming of fishes shelled eggs of reptiles and birds adaptations of wings, feet and beaks of birds viviparity of mammals homoiothermy of birds and mammals divergence, convergence and parallel evolution gliding and true flight of vertebrates general features of arthropods especially insects general features of molluscs insect pollination of flowers evolutionary series fossil record basic biochemical unity changes in basic biochemistry origin of life	9 12 20 24 25 31 34 39 42 42 46 48 49 51 51 53 54 55	history of the Earth geological record further theories about the origin of life early atmosphere of the Earth some aspects of human societies	27 27 27 27 33, 34

Objectives

When you have finished your study of this Unit, you should be able to:

1 Define correctly, or recognize the best definitions of, or distinguish between true and false statements concerning each of the terms and principles listed in column 3 of Table A. (Tested in *SAQs* 1, 4, 7, 8, 9, 11)

2 Given a list of sixteen organisms, specifically referred to in the course, identify them as vertebrates, insects or molluscs. (Tested in *SAQ* 5)

3 List or identify the primary advantages which accrue from each of the following:
homoiothermy, flight, viviparity, air-breathing, possession of limbs.
(Tested in *SAQs* 2, 12)

4 Associate correctly given innovations, e.g. lungs, homoiothermy, with appropriate changes in habit or habitat. (Tested in *SAQs* 2, 12)

5 List or identify at least two differences between birds, bats, and pterosaurs, taken as singletons or as a pair and a singleton. (Tested in 21.4. *SAQ* 12)

6 Given statements about flowers and insects, match those indicating co-adaptation. (Tested in *SAQs* 6, 12)

7 List at least three biochemical mechanisms, or substances connected with them, common to all living organisms and at least two mechanisms or substances not common to all—or, given the mechanisms or substances, classify them as either common to all or not common to all. (Tested in *SAQs* 7, 8, 9, 11)

8 Given a series of statements pertaining to adaptive radiation or the origin and evolution of organisms, indicate their scientific validity. (Tested in *SAQs* 4, 10, 13)

9 Draw accurate conclusions from given bodies of reliable data, state them clearly and concisely and give or select reasons for the conclusions. (Tested in Appendices 2 and 3)

10 Distinguish between and identify accurately teleological and non-teleological statements. (Tested in *SAQ* 3)

11 Formulate testable hypotheses from given evidence and data. (Tested in 21.1, Appendices 2 and 3, and *SAQ* 13)

12 Apply principles expanded in this and earlier Units to totally new (given) situations to which they are relevant. (Tested in Appendices 2 and 3, and *SAQ* 13)

Study comment The diversity of living organisms is impressive and can be bewildering. The intention of this Unit is to illustrate how acceptance of Darwin's theory of evolution through natural selection provides a key to understanding the diversity of life. This diversity of form and habit is based on remarkable similarity of basic biochemistry, as might be expected if all organisms living on Earth have a common origin. Some facts are necessary to illustrate the principles. You should not try to remember the facts, but it is important that you should understand the principles and how they are applied by biologists.

Student's Guide

This is the last of a block of 'biology Units'. Its main theme is aspects of the process of evolution that have led to existing organisms. Essentially it is based on Units 19 and 20, where you studied natural selection and factors which affect the survival of organisms; these in turn were based on Units 17 and 18. When we go back in time to consider the past history of organisms, we anticipate part of the next block of Units, devoted to the Earth sciences, and when we speculate about the origin of life we refer back to Units 14, 15 and 16 for biochemical evidence. The inter-relationships between these Units are shown in the Conceptual Diagram (p. 4).

In Unit 21, we survey very briefly the two largest groups of living animals, the arthropods and molluscs (defined in 21.5); and we study the next largest group, the vertebrates (defined in 21.3) in slightly greater detail. This is because we are ourselves vertebrates and other members of this group are familiar to us. Do not attempt to remember the names or characteristics of organisms or groups which you do not already know, but try to recognize the principles we illustrate from these organisms— the principles that underlie evolutionary biology. The Unit includes a table of geological periods (p. 23)—a list of somewhat formidable names and dates. You are not expected to remember any of these, but you will find it easier to follow section 21.3, on the evolution of the vertebrates, if you refer back to this table occasionally.

The introduction includes a list of four generalizations that are basic to evolutionary studies. You will attempt to put these into practice during your Home Experiment, when you sort out 'evolutionary lines' of nails, screws, paper clips and so on. You should bear these generalizations in mind as you read the rest of the text.

The idea of a species as an interbreeding population of organisms was developed in Unit 19. In section 21.1, we consider the evolution of new species, using a group of birds that were studied by Charles Darwin himself. The section is constructed as an exercise in which you should reach certain conclusions by study of data provided and by answering a linked series of questions. In the rest of the Unit, we are concerned with evolution of forms that are more distantly related to each other than are species, that is with the evolution of families, classes and higher groups of organisms. In section 21.2, we note that the diversity of modern mammals is based on a unity of pattern of many anatomical and physiological features, as though there had been modification in different ways of characteristics basic to the group and presumably present in their common ancestors. This is one example of 'adaptive radiation', the phenomenon that results from natural selection of variants in relation to particular features of the niches that these organisms come to occupy. Only organisms that are well adapted to the conditions under which they normally live are likely to survive to reproduce their genetic complement. Major groups evolve as a result of mutations and rearrangements of genes that make adaptive radiation possible. We can find evidence that ancestral mammals were different from present-day ones by looking for fossils (defined in 21.2). The fossil record is introduced in 21.2.1, but you are not expected to remember the names of Eras and periods nor the dates.

In section 21.3, we widen the scope of our discussion by considering the vertebrates. The section is a historical account of the evolution of the vertebrates, with comments on the adaptive significance of the structures evolved by each group; this is supported by study of the divergence shown among modern members of each group. Thinking of an organism as simply a collection of parts, each adapted to a particular habit, is helpful in many ways, but it can be dangerous unless you realize that the living organism must be an integrated whole. This is particularly stressed when discussing birds in section 21.3.4.

One of the consequences of adaptation to particular habits is that unrelated organisms occupying similar niches often resemble each other closely. This is the phenomenon of 'convergence', discussed in 21.4 and illustrated by a detailed comparison of flying vertebrates. The Unit's TV programme shows convergence among swimming vertebrates.

If you are having difficulty and are short of time, you should read rapidly through section 21.5, which introduces the two largest groups of living animals, the arthropods (which include the insects) and the molluscs. Section 21.5.1 refers briefly to flowers that have evolved in parallel with certain groups of insects.

Section 21.6 points out some of the difficulties of drawing conclusions from the fossil record; this section leads to a consideration of biochemical evidence for evolution in 21.7. All the information in this section should be familiar to you from Units 14 to 18, so you should be able to read quickly through it, revising your previous work by viewing it from a different angle.

The final section, 21.8, concerns speculations about the origin of life. You will meet other speculations about this in the Earth Sciences Units (22–27). One of your Home Experiments illustrates a type of observation that is relevant to this problem.

Appendix 1 is an account of some of the impressive fossil reptiles that are found in rocks aged between about 225 and 65 million years. The reasons for their extinction are still being debated. The best time to read this appendix, if you have the time and interest, would be after reading section 21.3.2.

Appendix 2 is an exercise to illustrate the features of convergence and divergence by study of the marsupial mammals of Australia and comparison between them and mammals in other parts of the world. The best time to do this exercise is probably after you have read sections 21.1 to 21.4, but you could do it later, as an *SAQ*, if you wish.

Appendix 3 is an exercise in which the genetic code, given in Unit 17, is used to rank the number of mutations needed to transform part of the protein cytochrome c of various organisms into the equivalent human protein. The exercise illustrates the sort of biochemical differences that are observed among organisms and also the way in which these differences can be used to deduce whether organisms are closely or distantly related to each other. The exercise fits best after section 21.7, but you could do it later as an *SAQ* if you wish.

There are two prescribed texts for this Unit:

The Chemistry of Life is, by now, very familiar to you. You are directed to read extracts from it in sections 21.7 and 21.8. The first of these extracts is largely revision of material treated in earlier Units.

Man and the Vertebrates (volume 1) by A. S. Romer is an account of the evolution of the vertebrates; it includes descriptions of the anatomy of some modern forms as well as information about fossil vertebrates. There are many diagrams and photographs and you should look through these as further illustrations for sections 21.2 to 21.4 of this Unit. You should read Chapter 9 (The Origin of Mammals) after you have finished reading the Unit; it illustrates the sort of contributions to understanding of modern groups that come from studying fossils and comparative anatomy. Read this for its general interest—do not try to remember the names, terms and facts. You can read the rest of the volume as equivalent to a black-page Appendix if you wish. You met volume 2 with Unit 19; it includes information about the evolution of man.

Introduction

In Unit 19, we examined the way in which living organisms can change with time—that is, how evolution can happen as the result of mutations (heritable variations) arising in populations of interbreeding organisms, which become adapted more and more closely to their environments. As a result of competition, both within and between populations, selection should result in changes of gene frequency; changes in climatic conditions and in food supply and in predation may also cause changes in gene frequency. Thus, populations which once consisted of similar organisms may, over a period of time, alter so that two or more dissimilar and non-interbreeding populations are produced. The unit of evolution is, by this analysis, the interbreeding population rather than the individuals that make up the population.

Since Charles Darwin and Alfred Russell Wallace first proposed their theory of evolution by natural selection in 1858, many biologists have worked in this field and have established certain generalizations:

1   Organisms that resemble each other in many ways are usually more closely related evolutionarily than organisms that resemble each other only slightly. This is a consequence of the origin of distinct species by the splitting up of an older species—the descendant species will share some of the features of their common ancestral species.

generalizations about evolutionary changes

2   Evolution proceeds by the gradual accumulation of small changes in structure and function. The mutations on which selection works are generally small alterations in the characters of the ancestor—a single amino-acid substitution in a polypeptide (as in sickle-cell haemoglobin, see Unit 19) rather than the sudden development of a new organ or organ system.

3   Simpler forms gave rise to more complex ones and smaller forms gave rise to larger organisms. There are certain difficulties in defining 'simple' and 'complex' and the two statements just quoted can only be applied in a very general way.

4   Evolutionary processes do not go into reverse. Larger, more complex organisms that have evolved from smaller, simpler ones do not subsequently evolve into smaller, simpler forms. If the larger, complex type is at a competitive disadvantage compared with a simpler, smaller type, then usually the larger, complex type becomes extinct—its genes vanish.

Stating the broad results of evolutionary studies in this way suggests that there is a body of facts that supports them—but this is not strictly true. There can be no reasonable doubt that species evolve by selection altering gene frequencies in interbreeding populations—no reasonable doubt because there is experimental evidence supporting the theory. But there is no such direct support for the application of this theory above the species level. All the evidence about the evolution of families, classes and larger groups of organisms is indirect and circumstantial—but, taken all together, the facts seem to be consistent with the theory of evolution by natural selection.

In much of this Unit, we shall be concerned with the phenomenon of *adaptive radiation*. This term describes the evolution of many species, adapted to a range of different niches (see Unit 20), from an ancestral species with a restricted niche which may have been quite different from any of those occupied by its descendants. Related organisms, living in different niches, will differ in certain features related to the problems for

adaptive radiation

9

survival in those niches. Each form descended from a common ancestor will survive only if it is adapted to the niche in which it lives; it will therefore differ from other forms descended from the same ancestor but living in (and adapted to) other niches. Adaptation to particular niches will include anatomical features, especially those related to feeding and movement, and physiological features, such as those related to breathing and breeding. But, since the descendant forms have a common ancestor, they are likely to share many features inherited from that ancestor—this is an application of the first generalization stated above. In practice, biologists tend to work backwards—they observe that two groups share common features not related to the special niches occupied by them, and from this they deduce that the two groups share a common ancestry.

The sections are arranged in the following order:

In 21.1, you are asked to use certain facts about the finches on the Galapagos Islands to reach conclusions about their probable evolution by answering a series of questions. Some of these finches were collected by Charles Darwin during his voyage round the world on HMS *Beagle*; his study of them and of other animals on the Galapagos Islands helped convince him that evolution had occurred through the operation of natural selection.

In 21.2, we extend the idea of divergence from a common ancestral type to the mammals. Looking for possible ancestors introduces the *fossil record*, based on earth scientists' ability to age rocks.

21.3 is an account of the evolutionary story of the vertebrates, surveying the modern groups by stressing their special adaptive features.

One consequence of adaptive radiation is the phenomenon of *convergence*, treated in section 21.4, using vertebrates as examples.

Section 21.5 describes two other large groups of animals, the arthropods and the molluscs; there is a sub-section on the mutual adaptations shown by certain insects and certain flowering plants.

In section 21.6 we return to the fossil record and some of its limitations, and work back to the earliest forms of life. Section 21.7 reminds you of the many biochemical mechanisms common to all living organisms, but points out that evolution also occurs at the biochemical level.

Finally, in 21.8 we speculate about the origin of life.

The Appendices are all black-page material. Appendix 1 is about the large fossil reptiles that became extinct more than 65 million years ago; you are advised to read this after section 21.3.2. Appendix 2 is an exercise based on the peculiar mammals of Australia; it illustrates the principles underlying sections 21.1 to 21.4, so it is best read after reading these. Appendix 3 is another exercise; differences in part of the cytochrome c molecule of certain organisms are used, with the genetic code, to illustrate unity and diversity at the biochemical level—this is best studied after 21.7.

21.1 The Galapagos Finches—a structured exercise

Instructions

In the exercise that follows we hope to do two things:

1 provide you with information about the Galapagos finches;

2 provide you with an opportunity to use certain facts to arrive at particular conclusions.

The exercise is written so that, after being given some information, you are asked to make a hypothesis or make a prediction based on the information you have just been given or on information given earlier. Some questions will require you to draw on your general knowledge or on your common sense or both. But you will not need special facts or information not given in the body of this exercise.

Answers to the questions are provided in section 21.1.1 (p. 17).

We suggest that you work through the exercise writing down your answers as you go and checking them against the answers we provide.

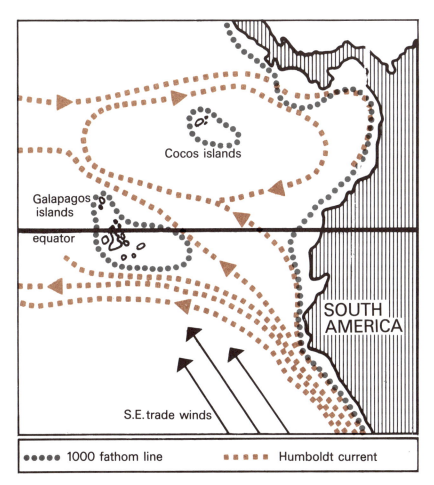

Figure 1   *Map of the Galapagos Islands and part of South America.*

The Galapagos Islands, which are volcanic in origin, lie on the Equator, 600 miles west of Ecuador and 1 000 miles south west of Panama. Being volcanic in origin, the islands were not inhabited by any living organisms when they emerged from the ocean, some two million years ago. Now the islands have plants and animals living on them. Although the islands in the group are all small (the largest is only about 80 miles long), a variety of plant communities is present. Mangrove swamps are found around some islands; the lowlands are barren, dry and populated by trees and bushes, while the higher land carries a wet tropical forest. The highest land of all has developed a grass/bracken community. Amongst the animals and particularly the birds on the Galapagos are to be found species that do not occur anywhere else in the world. Of these unique animals, a group of finches—thirteen different species—have been studied intensively. As they were first described scientifically in 1837 by Gould, who used specimens collected from the islands by Charles Darwin in 1835, the whole group came to be called Darwin's finches.

The nearest land with birds that even remotely resemble Darwin's finches is 600 miles away. So how can we explain the origin of these birds? Where could they have come from and how? The obvious place, Ecuador or Panama, is well beyond the usual distance that small birds fly; but strong tradewinds blow from mainland America towards the Galapagos, so it is possible that colonization of the islands occurred from the mainland.

Question 1

Bearing in mind that 600 miles separates the islands from Ecuador do you think that colonization from the mainland would be a frequent or a rare event?

Question 2

Suppose that colonization from the mainland occurred, would you expect there to be a resemblance between mainland and island finches?

Question 3

As the island finches are unique, although clearly finches, could they have arrived like this in the past?

Assuming that the islands were colonized from the mainland we can suppose that this happened in one of two ways, shown in Figure 2.

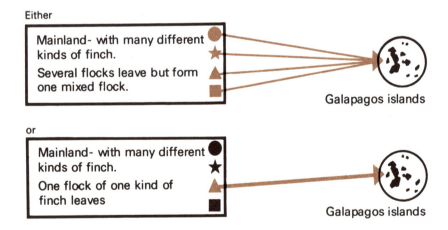

Figure 2 *Diagram to compare two possible theories of the origin of Galapagos finches from South American finches.*

Examine the drawings of the finches in Figure 3.

Question 4

Are the differences between the finches, the sort that you would expect if they were all closely related to each other? Are these differences of the kind that you would expect if the finches had colonized the islands as a mixed flock from the mainland?

12

Large ground finch

Medium ground finch

Small ground finch

Sharp beaked ground finch

Cactus ground finch

Large cactus ground finch

Large insectivorous tree finch

Vegetarian tree finch

Woodpecker finch

Warbler finch

Large insectivorous tree finch on Charles

Cocos finch

Small insectivorous tree finch

Mangrove finch

Figure 3 *Males of each of the species of Galapagos finches and of the Cocos finch, all drawn to the same scale: all are drab in colour.*

13

Examine the map of the islands (Fig. 4) and then look at Table 1.

Table 1

The percentage of finch forms which are peculiar to each island (or pair of islands) and not found anywhere else.

Islands near the edge of the group	Percentage	Percentage	Islands near the centre of the group
Culpepper and Wenman	75	0	Indefatigable
Tower	67	5	James
Hood	67	14	Barrington
Abingdon and Bindloe	33	20	Albemarle and Narborough
Chatham	36	25	Charles

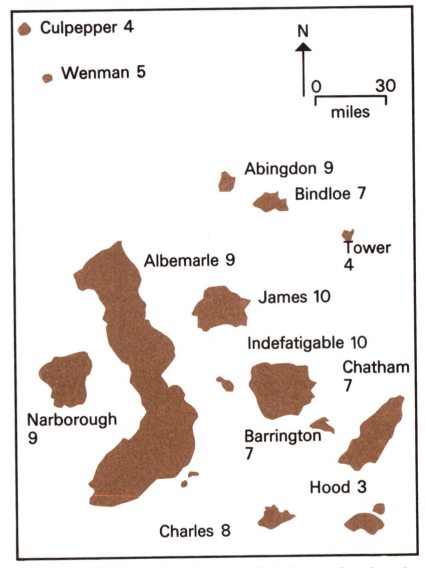

Figure 4 *Map of Galapagos Islands. The numbers after each name indicate the number of finch species that occur on that island.*

Notice (1) that the islands at the edges of the whole group have fewer species of finch than do the central islands, but (2) that the proportion of finches in the total population of each island that is peculiar to that individual island, (i.e. is endemic) is higher in the islands at the edges than amongst the islands in the centre.

Question 5

Suppose that the islands were colonized by a mixed flock of finches from the mainland, is the distribution of endemic species the one you would expect?

Question 6

Suppose that a flock of finches all of one species colonized the islands, what might have happened that could account for the distribution of endemic species that is observed today?

Finches are small birds and are not powerful fliers, nor do they fly for long distances. If a flock of birds all the same species had colonized the island in the distant past they might be expected to fly short distances but not long ones.

Question 7

Suggest one explanation that would explain both (a) the high proportion of endemic species on Culpepper, Tower and Hood, and (b) the fact that islands close together have finch species in common.

Darwin's finches, taken as a whole, feed on both animal and vegetable food material—seeds, buds, leaves, fruit, insects of various kinds and sizes, and nectar from flowers.

Examine the drawings of the skulls of two of the finches in Figure 5.

Question 8

Which finch of the two shown is most likely to feed on small insects?

Question 9

Of the items in the food list above, which do you think would be the main food of *G. magnirostris*?

Question 10

Now suppose both the finches *G. magnirostris* and *C. olivacea* ate similar types of food, how would the beak size and shape affect the birds' diet? If they both ate insects, would they both be able to eat large hard-bodied insects? If, on the other hand, they both ate seeds, would they both be able to feed on the same kind of seeds?

Question 11

Condition (a): if all the birds on an island have beaks of the same size and shape one would expect them to be feeding on similar foods, e.g. seeds or insects.

Condition (b): if the beaks differ in size and shape one would expect the birds to feed on different types of food, e.g. seeds of different kinds and/or insects of different kinds.

> **Under which condition (a) or (b), would you expect:**
> 1 **there to be more birds on the island;**
> 2 **parent birds to be able to feed and thus to rear to maturity more offspring?**

Amongst Darwin's finches there is a group of six species that resemble each other enough to be put into the same genus, *Geospiza*. The species in this genus are called the ground finches—finding their food mainly on the ground and in plants near the ground. When the beaks of this group of finches are examined, a variety of differences in beak design are found. A similar sort of variety is found amongst species of finch feeding on insects and buds and leaves. Some of the differences in size are shown in

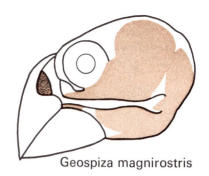

Geospiza magnirostris

Certhidea olivacea

Figure 5   The skulls of two species of Galapagos finch, drawn to the same scale. The jaw muscles are coloured.

15

Table 2. Figure 6 shows the shapes of their beaks. Not all species of finch are found on each of the islands. Consider the ground finches on the islands of Tower, Hood, Culpepper and Abingdon. Their heads are shown in Figure 6.

Table 2

Beak sizes in mm

After birds reach maturity their beaks do not grow any more, so the sizes shown are average sizes for mature birds. Of the two figures under each name, the first is the length and the second is the depth of the beak.

Ecological type of finch	Species found on:			
	Tower	**Hood**	**Culpepper**	**Abingdon**
Large ground finch	*G. magnirostris* 16.5, 21.1	*G. conirostris* 15.4, 16.0	*G. conirostris* 15.0, 16.5	*G. magnirostris* 16.0, 20.0
Medium ground finch	*G. conirostris* 14.4, 13.0			*G. scandens* 14.6, 9.7
Small ground finch	*G. difficilis* 9.4, 7.9		*G. difficilis* 11.3, 9.0	*G. difficilis* 9.7, 8.5

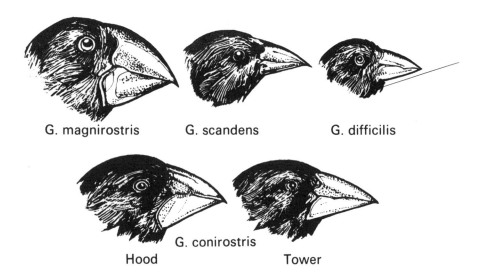

G. magnirostris   G. scandens   G. difficilis

G. conirostris
Hood   Tower

Figure 6   Diagrams of the heads of five types of Galapagos finches drawn in profile to compare the shape and size of the beaks.

Question 12

Consider the data shown in the table for the large and medium ground finches. Suggest an explanation, in terms of food, for the differences in average beak size of *G. conirostris* on Tower and Culpepper.

Notice that *G. conirostris* is absent from the island of Abingdon where *G. scandens* is present.

Question 13

Suppose that it were possible to catch a large number of *G. conirostris* on Tower and to release them on Abingdon. What effect do you think that the imported birds would have on the population of *G. scandens* already there?

16

21.1.1 Answers to the questions on the Galapagos finches

Question 1

A rare event rather than a frequent event.

Question 2

Yes. There would probably be strong resemblances.

Question 3

No, at least not in the recent past.

Question 4

It is a matter of common observation that animals related to each other are more like each other than they are like unrelated animals of the same type. So in this case all the finches being similar in shape and size and in colour suggests a family-type relationship. If the original colonization had been a mixed flock of finches, it would be reasonable to expect to find the present-day finches more unlike each other than in fact they are.

Question 5

If the original colonizing flock was a flock of several kinds of finch, then it would seem reasonable to expect that each island would have the same number of species today. On this hypothesis, the present-day distribution is not the one expected.

Question 6

If the colonizing flock was all of one species, then the other species now present in the islands must have evolved on the islands.

Question 7

The high proportion of endemic species could be explained by supposing that it was on these islands that the particular endemic species evolved: because these islands are isolated from the others, the finches did not spread out and colonize other islands. In the case of islands close together, the finch populations were able to cross from island to island, and so no island developed its own endemic species.

Question 8

Certhidea olivacea.

Question 9

Seeds probably and large insects. The beak is a massive one and looks as if it would be capable of cracking open seeds and dealing with the larger heavy insects.

Question 10

It seems likely that birds with beaks as different as the beaks of these two species would in fact eat different types of food. The heavy beak of *G. magnirostris* does not look at all suitable for picking up small delicate food, whether insects or seeds.

Question 11.1

Given that the total number of birds is limited only by the amount of food available to them:

if (a) . . .

then all the birds would be exploiting similar food material, and other types of food material would be unused by them;

if (b) . . .

then all the food materials of all types might be exploited by different kinds of finch. In this case, as food materials of different kinds are being used, more birds would find enough food to eat.

Question 11.2

The ability of parent birds to feed and rear young is limited by the supply of food suitable for the young (which may not be the same as that suitable for the adults). Again, it is likely that in condition (b) more birds would be successful in rearing young.

Question 12

Both *G. magnirostris* and *G. conirostris* live and breed on Tower, but *G. conirostris* is the only medium/large ground finch on Culpepper. So presumably the food available to it on Culpepper includes not only the food it eats on Tower but also the food that *G. magnirostris* would eat if it were present. To deal with this food, larger and heavier beaks might be an advantage. But on Tower the individuals of *G. conirostris* with larger heavier beaks would have to compete with individuals of the other species feeding and living in this ecological niche.

Question 13

Presumably the introduced birds and the native birds would compete for the same types of food. This might affect the breeding success of one or other of the two species, or perhaps both might be affected. If one species for any reason were more efficient at collecting food, then the other species would, over a period of generations, die out. If both species were equally efficient, then it might be that both species would continue to live and breed on the island, but in reduced numbers.

21.1.2 Conclusions

Answers to these questions and other questions of similar type enable us to put forward an explanation of the distribution of Darwin's finches and of the differences between them.

The islands were colonized initially by finches from mainland South America—possibly blown, or at least assisted, by wind. In view of the isolation of the islands, colonization probably happened only a few times, perhaps once only.

As the islands were uninhabited, as far as finches were concerned, and there were probably no other small birds on them, the colonizing population was able to feed and reproduce, and the finch population expanded. Owing to the wide range of food available, and to the vacant feeding and breeding niches, the finches were able to exploit vacant ecological niches. Those finches that developed characters enabling them to exploit their niches most effectively bred successfully—they found plenty of food and hence the broods of young birds could be reared safely. At the same time, removal of pressure on the resources of the environment meant that the unchanged finches were also more successful. Hybrids between the varieties were possibly less well equipped than either parent to exploit niches and were consequently less successful at breeding.

Distant islands were possibly colonized by finches from the main group of islands, which, being closer together, shared the same finch fauna. Probably not all varieties of the finches arrived on the distant islands; or, if they did arrive, some may have found their niche already occupied— they would have had to compete for food and other resources with resident finch populations.

So the more distant islands have more endemic species of finch than the islands closer together in the central group.

Of course, this explanation of the present state of affairs, while it fits our observations, also contains a number of assumptions and speculations. We cannot know for certain the course of evolution in Darwin's Finches, but can only put forward a more or less plausible guess.

Charles Darwin himself wrote (in *The Voyage of the Beagle*):

> I never dreamed that islands about fifty or sixty miles apart, and most of them in sight of each other, formed of precisely the same rocks, placed under a quite similar climate, rising to nearly equal height, would have been differently tenanted. It is the circumstance, that several of the islands possess their own species of the tortoise, mocking-thrush, finches and numerous plants, these species having the same general habits, occupying analogous situations, and obviously filling the same place in the natural economy of the archipelago, that strikes me with wonder.

21.2 Adaptive Radiation in Mammals

The study of Darwin's finches (the Galapagos finches) shows how adaptive radiation, when combined with isolation, can lead to the evolution of many species from a single group of ancestors. Let us see whether the idea of the evolution of new types by adaptive radiation can be applied above the level of genus and species.

These mammals (Fig. 7) obviously differ in bodily proportions and in habit: bats fly; whales and seals swim; elephants and hippos are ponderous and move slowly; giraffes and horses are fleet-footed herbivores; dogs, cats, lions and shrews are carnivores; the mole is a digging (fossorial) carnivore; lemurs and squirrels are arboreal forms—and you will probably think of other mammals with different habits.

In spite of the differences between all these animals, they have certain similarities in structure and physiology and these are so great that we must conclude that all mammals are more like each other than they are to other types of animals. (Some of these similarities are mentioned in 21.2.1 and in 21.3.5.) We therefore presume that present-day mammals are the product of a process of evolutionary divergence similar to that postulated for the finches of the Galapagos Islands. If this is true, then if we could observe mammal faunas of the past, we might expect to find less divergence than among the modern mammals. If mammals, in their turn, evolved from some other type of animal, then we might expect to find mammals with some of the characters of these ancestors and fewer of the mammalian characters with which we are familiar. It happens that we can examine animals of past ages to a limited extent by looking at structures called *fossils*. These are bits, whole bodies or casts of organisms, preserved as a result of a series of chemical changes in the rocks.

fossils

21.2.1 The fossil record

In Unit 27, you will learn about the history of the Earth and about the methods used for measuring the age of rocks. Here we shall assume that the age of certain rocks can be estimated with reasonable accuracy, so that we know the approximate dates at which the structures within them were formed. We can compare rocks from different parts of the world and arrange them in chronological order.

From studying certain features of the rocks and of the fossils in them, we can deduce the type of environment which existed at the time when the fossils were formed. We can deduce whether the rocks were formed in the sea or in freshwaters or on land, and whether the climatic conditions were humid or desert or showed the sort of annual fluctuations which we experience in the British Isles today. So we may have a surprisingly clear mental picture of conditions on Earth many million years ago.

There are, however, gaps in the time sequence and in the range of habitats and climates in which fossils were formed at different times. The finding of a fossil is positive evidence that the organism existed, and it may be possible to describe it fairly completely, but a gap in the record merely means that we have failed to find fossils of a certain age or certain kind. They may or may not have existed. Sometimes the discovery of a new area with fossils in its rocks leads to a re-thinking of previously held ideas.

Flying insectivore

Omnivore

Arboreal

Large terrestrial herbivore

Amphibious herbivore

Fossorial insectivore

Carnivore

Aquatic piscivore

Aquatic filter feeder

Figure 7 A selection of modern mammals, not drawn to scale.

You will probably find fossils in your local museum. There is a very good display in the Natural History Museum in South Kensington, London.

We recognize modern mammals from their 'soft parts'—the hair or fur on their bodies and the mammary glands through which they suckle their young are two important features. But fossils are usually hard parts of organisms—shells, bones, teeth—so we have to use 'hard parts' to identify the animals. In the case of mammals, we look for certain features of the mammalian features chewing (molar) teeth and in the structure of the lower jaw and ear. If these are present, then we call the fossil a mammal; if they are absent, then it is not a mammal. Using these criteria, some fossils in rocks formed about 190 million years ago are the earliest known mammals, but the earliest specimens assigned to the group of modern mammals are about 80 million years old.

There is sufficient fossil material of mammals for us to be certain that the present mammalian fauna does not include all the types of mammals evolved in the past. We lack a detailed knowledge of the precise course of evolution of the mammals alive today. For instance, we are not certain when the whales evolved into a distinct group within the mammals, but there is a fossil whale about 50 million years old. What is certain is that, after the modern mammals became distinct as a breeding group, adaptive radiation followed, resulting in a varied collection of animals which nevertheless retain many fundamental similarities. The development of diversity of form among the modern mammals has been imposed on a basic pattern of mammalian organization; this has been retained, we believe, from early mammals, and perpetuated to modern forms.

Table 3

Table of Geological Periods (omitting older Eras, before fossils became abundant)

Era (duration years × 10⁶)	Period	Estimated time since beginning of period (× 10⁶ years)	Life
Cenozoic Age of mammals (66)	Quaternary	1.5	Modern species of mammals; dominated by man.
	Tertiary	65	Radiation of placental mammals and birds; third radiation of insects.
Mesozoic Age of reptiles (160)	Cretaceous	136	Flowering plants begin to radiate; large reptiles extinct by end of period.
	Jurassic	190	Reptiles dominant on land; first birds; archaic mammals present; first teleosts at end of period; second radiation of insects.
	Triassic	225	First dinosaurs, ichthyosaurs, plesiosaurs, turtles. Conifers dominant on land. First mammals at end of period.
Palaeozoic Age of fishes and animals without backbones (345)	Permian	280	Reptiles begin to radiate and replace amphibians as dominant vertebrates on land. Ancestors of teleosts radiate in sea.
	Carboniferous	345	Coal forests of ferns and mosses; radiation of amphibia; first reptiles; swamp pools full of bony fishes; many sharks in sea. First radiation of insects.
	Devonian	395	Age of fishes (mostly freshwater); first large land plants; first amphibians.
	Silurian	440	First land plants and first insects; primitive vertebrates (mainly freshwater).
	Ordovician	500	First vertebrate fossils; lamp-shells and molluscs (cephalopods) dominant in sea.
	Cambrian	570	All main types of invertebrates present in sea; diverse types of algae (simple plants) in sea.

Use this Table as a guide to the main events that occurred during the evolution of modern plants and animals. The Earth Sciences Reader, *Understanding the Earth*, includes a large chart giving more information than this. You will refer to it while studying later Units.

DO NOT TRY *now to learn the names of the Eras and Periods, nor try to remember the times and details of fauna and flora.*

21.3 Adaptive Radiation and the Evolution of the Vertebrates

From examining a single class of animals, the Mammals, let us now extend
our discussion of adaptive radiation and evolution to a larger group, the
Vertebrates, which includes the mammals and also other animals that
resemble them in some ways but are very different in other ways. In the
vertebrates, as among the mammals, there is diversity of type imposed
upon a basic pattern of similarity in structure. The similarities include the
presence of a supporting skeleton within the body; a pumping heart
forcing blood rapidly round the body; and, at least in early stages of

Figure 8   *Representative modern vertebrates, not drawn to scale.*

development, a series of slit-like openings between the gut in the throat
region and the outside of the body. Vertebrates also typically have a tail,
a part of the body which does not include within it any part of the gut.
All these features of body structure are found only in vertebrates. Since
they share these features, fishes, frogs, snakes, birds and mammals are all
included in a single group, the Vertebrata. In grouping them together, we
imply that there is a genetic relationship between them—that they have
all evolved by divergence from a single ancestral breeding group.

The ancestors of the vertebrates probably lived on the bottom of the sea, and it is agreed that the first important step in their evolution was the appearance of animals with a supporting rod, the *notochord*, in the body, and with muscles divided into blocks arranged along the body on either side of it. These muscles contracted one after another, first on one side and then on the other, throwing the body into a series of waves (see Fig. 9). As these waves passed backwards along the body, so the body moved forwards through the water—that is to say, the animal swam.

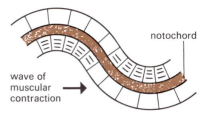

Figure 9   *Diagram to show notochord, with segmented muscles on either side of it.*

What selective advantages might this give the animal over its bottom-living ancestors?

The innovation that made swimming of the vertebrate type possible was followed by a variety of modifications of the body which included the following:

A swimming animal can escape from all predators except those that swim faster. It can also reach food and shelter by swimming towards suitable places.

(a) the appearance of a vertebral column made up of separate verte-brae. This replaced the notochord and made the body not only more flexible but also stronger.

(b) stream-lining of the body—this reduced the frictional resistance and so the amount of energy used.

(c) the appearance of fins at various places on the body—these improved swimming ability, allowing increased speed and control of the path of movement.

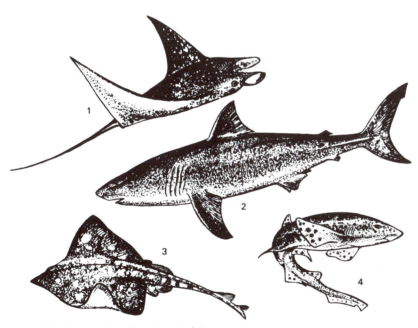

Figure 10   *Four modern elasmobranch fishes.*

21.3.1   Fishes

So radiation from the original vertebrate ancestor resulted in a great variety of fishes. These can be studied as 'swimming machines', in the same way that an aeronautics engineer studies the design of aircraft. This approach shows that the different shapes of fish and arrangements of fins are adaptations which allow different ways of life. Compare the four fish in Figure 10.

25

1 The manta ray flaps its large fins like the wings of a bird and swims constantly, feeding by straining small drifting animals (plankton) out of the surface waters of the ocean.

2 The great white shark is a very fast swimmer which feeds on large, actively swimming animals.

3 A small shark is a moderately active swimmer feeding on small animals on or near the sea floor.

4 A skate rests on the bottom of sandy seas and feeds on inactive animals.

All these fish live in the sea and breathe oxygen dissolved in the water through special organs called gills. None of these fish can live for more than a short time out of water, or in water containing little or no dissolved oxygen. They all have a flexible vertebral column, but the skeleton is not made of bone but of cartilage (familiar to you as 'gristle' in stews of meat or chicken).

The fish in Figure 10, the sharks and rays, are called the *elasmobranch* fishes (from the Greek words for 'plate' and 'gills'). Most of the other fishes have skeletons made of bone. Many species of elasmobranch fishes are widely distributed in the oceans, but they are seldom present in large numbers. They make up about four per cent of all the living species of fishes.

elasmobranch fishes

The earliest traces of vertebrates as fossils are bony plates in rocks about 460 million years old. Fossil skeletons of fishes become more common in rocks about 390 million years old, and they are very numerous in rocks slightly younger than this. The probable ancestors of the modern sharks and rays are present in large numbers in marine deposits (rocks laid down probably in the sea) aged about 390 to 300 million years.

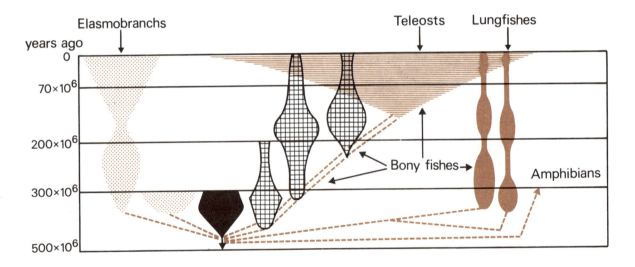

Figure 11 *Diagram to show the distribution of fossil fishes in rocks of different ages. The width of each part indicates the number of fossils of that group found in rocks of that age. The oldest rocks are at the bottom of the diagram.*

Most of the early fish remains are in freshwater deposits, probably laid down in a hot climate with marked seasonal droughts. These sorts of conditions are found today in tropical swamps that may dry up. The abundant decaying vegetation often uses up all the oxygen in the water during wet periods, so animals living in these swamps are exposed to lack of dissolved oxygen during both wet and dry seasons. The fish living in such waters today obtain the oxygen, which is essential for their respiration, from the air rather than from the water.

The ancient fishes that lived in tropical swamps almost certainly could breathe in air as well as in water. They used gills when breathing in water, and for air breathing they used a pair of lungs. There is a small group of modern fishes called 'lungfishes' (see Fig. 12) which behave in exactly this way, and we believe that their breathing arrangements have been modified very little from those of their remote ancestors. We shall return to them later.

lungfishes

Figure 12 *The Australian lungfish* Neoceratodus.

Some of the ancient fishes moved from swamps into the sea, and fossils of this group are very common in marine deposits about 225 million years old. Presumably these fish were able to rely entirely on gill respiration and no longer used their lungs. From studies of the anatomy and development of modern fishes, we conclude that the lungs became transformed into a single structure, the *air-bladder*. You can see the position of this in the dissected trout shown in Figure 13.

air-bladder

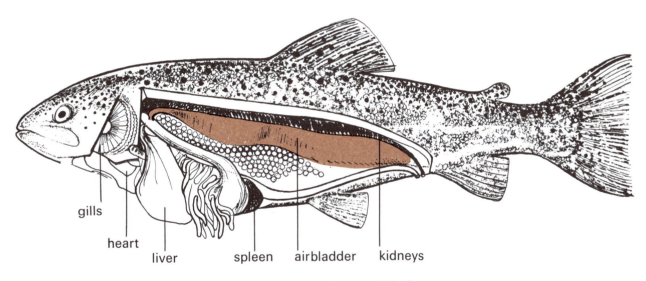

gills

heart

liver

spleen   airbladder   kidneys

Figure 13 *A rainbow trout dissected to show the internal organs. Note especially the air-bladder.*

This air-bladder has no respiratory function. Think of a submarine, and then try to suggest another function for it.

It provides the fish with controlled buoyancy.

Most of the tissues of the body are heavier than the same volume of water so the whole body sinks in water. Dogfishes and all other elasmobranchs lack an air-bladder. The dogfish can swim at any level or depth in the water, but if it stops swimming, it glides to the bottom and sinks because the body is heavier than water.

A fish with an air-bladder has built-in buoyancy, which it can alter so that it is able to float at any level in the water. You can observe this when you watch fish in an aquarium.

Perhaps they evolved from an ancestor earlier than the first fishes that had lungs (and lived in freshwater swamps). This conclusion is supported by the fossil record, as illustrated in Figure 11.

The fishes which have evolved this sort of air-bladder and have thus achieved control of bouyancy are called *teleosts* (from the Greek for 'entire' and 'bone', because they also have bony skeletons, in contrast to the cartilaginous skeletons of the skate and dogfish).

teleosts

Within the teleosts there have evolved a wide variety of swimming habits, some of which are illustrated in Figure 14.

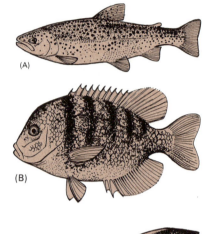

A   the trout—can swim fast and can twist and turn in the water as well as hover in midwater.

B   a fish related to our perch—swims more slowly than the trout, but it can swim backwards as well as forwards by using the long fins on its back and belly, and it can paddle itself using the paired fins.

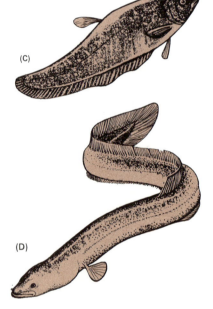

C   a 'featherback'—swims keeping its body straight and rigid and passing waves along its long fin below the belly. It can swim backwards by reversing the direction of this wave.

D   the freshwater eel—swims by throwing its whole body into a series of waves. It can also move on land, through damp grass, using the same body waves just like a snake.

E   a flatfish—actually swims on its side, using waves passing along its flattened body. It spends much time resting on the bottom on one side, and both its eyes are on the same side of the head (which is uppermost when the fish rests or swims).

You will see some fish swimming in this Unit's TV programme.

Curiously enough, the property which has opened up the largest number of niches is the ability to swim slowly, and to do so both forwards and backwards. Fish which can do this usually have long unpaired fins and pass undulations along these (see Fig. 14, B and C); they may use the paired fins as well, sculling forwards and backwards with these.

Figure 14   *Five modern teleost fishes illustrating different ways of swimming.*

28

The ancestors of the teleosts became common in the sea about 225 million years ago, and there followed an impressive evolution and radiation of types in the sea. Fossils with the diagnostic (special) characters of teleosts are found in rocks about 150 million years old, and rocks formed in the next 50 million years show an increasing diversity of types of teleosts (see Fig. 11).

About 95 per cent of modern species of fish are teleosts. In fact, there are more than 20 000 species and they occupy almost every imaginable aquatic habitat, exhibiting great diversity of form. They show remarkable physiological adaptation as well as anatomical variety. Species of teleosts live in such diverse habitats as the freezing waters of the Arctic and Antarctic Oceans; hot, mineral springs in Africa and Colorado; rushing torrents in the Andes and Himalayas; the perpetual cold and darkness of ocean abysses and of subterranean caverns. Some can even spend a lot of time on land. Of course, the majority live in less extreme environments in the sea and in freshwaters—but no teleost fish can live its life entirely out of water.

Some fishes have very bizarre shapes and various people have tried to describe them in mathematical terms as derivations of the 'ordinary' fish shape. Perhaps the most interesting attempt was that of D'Arcy Thompson. Those of you with sufficient time can read his own account in Chapter IX of *On Growth and Form* (abridged by J. T. Bonner). He used 'Cartesian transformation'—that is, he drew a regular network of lines, forming a grid of equal rectangles (see Fig. 15a), through the simple shape, and then drew a network of lines through equivalent points on the other shapes. Often, particularly if the fishes are closely related, the second network is a fairly simple transformation of the original rectangular grid (Fig. 15).

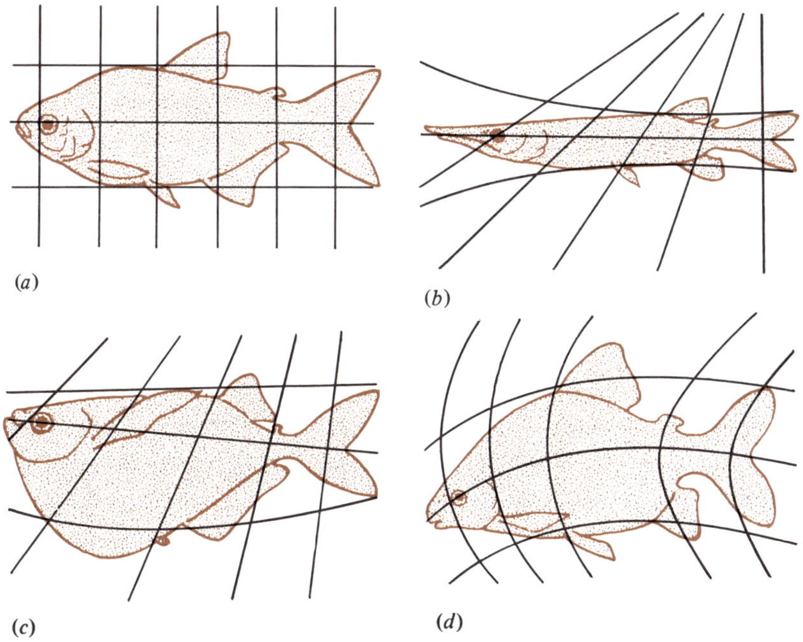

They can perform very precise feeding movements, bringing the mouth into contact with small objects and then giving a backwards pull or tug – very useful for pulling a fanworm out of its tube or sucking up the mud at the bottom of a pond or tugging leaves off plants. A fish that can swim slowly can manoeuvre itself into crevices through narrow openings, and it can thus escape from a predator that can only swim fast – such as a shark.

(a)

(b)

(c)

(d)

Figure 15   *Diagrams to illustrate how the shapes of related groups of fishes can be compared using Cartesian co-ordinates. The Characid (a) could resemble closely the ancestor of the three other groups (b, c, d).*

D'Arcy Thompson suggested that these cases show 'that variation has proceeded on definite and orderly lines, that a comprehensive "law of growth" has pervaded the whole structure in its integrity, and that some more or less simple and recognizable system of forces has been in control'. These forces remain to be investigated.

21.3.2 Terrestrial vertebrates

Another line of descent from these ancestral fishes that lived in tropical swamps led to terrestrial vertebrates—animals which may be wholly independent of water except in their diet. The first innovation in this direction was the evolution of paired limbs from the paired fins. In the ancestors, these paired fins consisted of a muscular lobe with a web round it supported by fin rays. The modern Australian lungfish has similar fins, but with a longer central lobe (Fig. 12). The African and South American lungfishes have lengthened the central axis and reduced or lost the web. They use these fins for propping up the body when resting on the bottom, and can push themselves forwards by bending the body and using the fins as struts (Fig. 16).

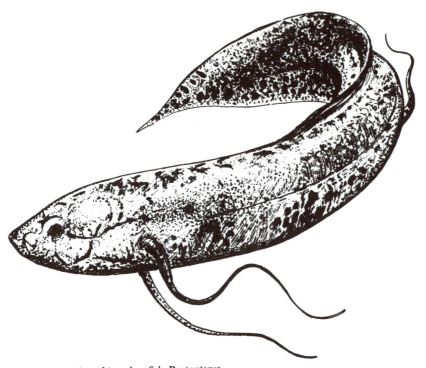

Figure 16 *The African lungfish*, Protopterus.

The earliest limbs had a short central axis, and developed a hand or foot with fingers or toes instead of the web of the fin. These limbs were probably used in the same way as the paired fins of lungfishes, but would have been more efficient struts. Vertebrates that have limbs cannot be classed as fishes; so the first to evolve limbs from fins must be classed as the earliest members of the *Amphibia*. The oldest amphibian fossils are about 350 million years old. These animals probably lived in the tropical swamps with their fishy relations and ate aquatic food and laid their eggs in water. They swam using their tails. Probably they breathed air as the modern lungfishes do, but some also had gills and could breathe oxygen dissolved in water—again as the lungfish still can.

amphibians

What advantage might they gain from having limbs instead of fins, although they lived in pools?

Very early in the evolution of vertebrates with limbs, some of them probably changed their reproductive habits and began to lay a new sort of egg. This would have been like a hen's egg (Fig. 17), but with a flexible leathery shell instead of the hard chalky shell.

When the pools in which they lived dried up, they could move overland to other pools. Since the contemporaneous fishes could have moved on land only as well as eels do today, the early amphibians might have had a better chance of reaching new pools and so of surviving and breeding.

When the embryo is very small, it floats on the yolk and it can be seen in a fertile egg as a small white disc against the yellow yolk. The embryo grows, using the yolk and the white of the egg as sources of food and water. Oxygen passes in through the porous egg shell and is absorbed by the embryo from the air space at the blunt end. The chick which breaks out of the egg is clearly similar in structure to the adult hen which laid it. Contrast these observations with the frog's life history (Fig. 18).

The frog lays eggs in water. Each is surrounded by a layer of jelly, and the whole lot stick together to form the familiar 'frog spawn'. The embryo develops using the small quantity of yolk, and it breaks out of the jelly (which dissolves away) as a small tadpole. This feeds in the water and breathes oxygen dissolved in the water, using gills. At first it swims using a tail.

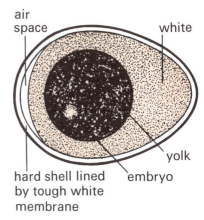

Figure 17   *Diagram of a hen's egg sliced along its length.*

Figure 18   *Stages in the life history of the common frog.*

Later, the tadpole grows legs and lungs. It starts to breathe air and soon it loses the tail and emerges on land as an adult frog, hopping with its legs and feeding on terrestrial animals. Typical amphibians resemble the frog in that they lay their eggs in water, and these eggs hatch into tadpoles with gills, breathing oxygen dissolved in water and swimming using their tails. So the amphibians typically are dependent on water for the development of their eggs and young. Contrast this with the hen: the hen's egg is laid on land and can develop with no water near it; the young chick is fully terrestrial, needing water only for drinking.

The first vertebrates to lay eggs that could develop entirely independently of water (i.e. eggs similar to those of the hen) were the early members of the class Reptilia. Fossils with diagnostic characters of reptiles are first found in rocks about 340 million years old. Like the amphibians, many of these lived in the water of the tropical swamps and fed on aquatic animals such as the fishes. But they breathed with lungs—no reptile ever develops gills. Like the amphibians, these early reptiles could move efficiently from pool to pool, and so probably had an advantage over the

reptiles

31

fishes when pools dried up. What the advantage could be of having eggs that developed out of and independently of water has puzzled biologists, but the answer is probably that these eggs were safe from the predatory animals which lived in the water. At that time there were few predators living on land, but the swampy pools were crowded with fishes, amphibians and reptiles many of which were probably predators ready to eat eggs and tadpoles. Reptile eggs were safe, and reptile young entered the water when larger than amphibian tadpoles (because the eggs contained more yolk) and so were more likely to survive. Thus the laying of terrestrial eggs was actually an advantage to animals which spent most of their lives living and feeding in water.

Name modern reptiles which lead similar lives.

Turtles and crocodiles live and feed in water as adults, but lay their eggs on land. There are a few species of lizards and snakes which also do this.

But in the modern world there are many terrestrial predators; turtles and crocodiles are in grave danger of extinction in many places, at least partly because of very high predation of their eggs and their young before they reach the comparative safety of the water.

Reptiles all breathe through lungs and lay their eggs on land. The early ones had limbs suitable for walking on land. So it is not surprising that fully terrestrial reptiles soon evolved, and showed a remarkable adaptive radiation.

Name some modern reptiles which are fully terrestrial.

Most lizards and snakes; the tortoises (relatives of the aquatic turtles).

Turn to Appendix 1 (Black) to read about some of the variety of fossil reptiles that are found in rocks between 225 and 65 million years old—animals that lived during the 'Age of Reptiles'.

21.3.3 Homoiothermic vertebrates

All modern reptiles, like the fishes and amphibians, have body temperatures which vary with that of the environment: i.e. they are *poikilotherms*. When the air around them is warm, then they are warm and active, but if the air is cold, then they become chilled and lethargic. We discussed the possible advantages of *homoiothermy* (of being able to maintain a constant body temperature) in Unit 18 (18.4.3). Homoiotherms are able to remain active over a wide range of external temperatures and therefore have a greater chance of survival in most parts of the world than have poikilotherms.

Two groups of reptiles evolved adaptations which allowed them to become homoiothermic, and these gave rise to the birds and mammals. We cannot be certain whether or not some of the extinct reptilian groups, e.g. dinosaurs and pterosaurs, had become homoiothermic. But the independent development of homoiothermy in birds and in mammals is a good example of parallel evolution of a physiological system. Many of the details of the structures evolved in the two groups are different as we can demonstrate by comparing body temperature control in birds with that in man. Refer back to Unit 18, section 18.4.3 if you cannot answer the next three questions.

What are the important structures concerned with heat loss in man?

Capillaries which bring warm blood to the skin; the skin from which heat can be lost by radiation; sweat glands that secrete a watery fluid which is evaporated and so cools the skin and the blood in the capillaries.

Some other mammals have sweat glands and a good capillary supply to the skin. A few, such as a dog, lack sweat glands. A dog loses heat by panting. Much water is evaporated from the tongue, mouth and lungs, leading to cooling of the rich blood supply to these organs. Birds lack sweat glands and they lose heat in the same way as the dog—that is, they open their mouths and pant.

What are the structures concerned with heat production in man?

Muscular contractions provide heat, especially during shivering. Most of the chemical reactions within the body result in some heat production.

Some mammals (e.g. bears, hedgehogs) *hibernate*—that is, they pass the winter in a lethargic state when the blood temperature falls and they appear to be asleep (or dead). They revive again in the spring. A few bird species also hibernate, for example the poor-will, a relation of the nightjar, in California. Hummingbirds at high altitudes become torpid and cold each night and revive and become active again when the sun rises next day.

What structures provide insulation in man?

The fat deposits under the skin are the most important system of insulation.

Man has a sparse covering of hair. Other mammals have more hair (or fur) and this usually provides excellent heat insulation. Whales and dolphins are almost hairless; they have thick fat deposits (blubber) under the skin. Some birds also have thick fat deposits under the skin. In addition birds have feathers which provide excellent insulation. Like a covering of fur, a layer of feathers retains motionless air immediately outside the skin, and this greatly reduces heat loss. You may have observed small birds such as robins looking remarkably plump in cold weather—how can you explain this? There are muscles which raise the feathers; this has the effect of deepening the layer of still air, and so reducing heat loss even further. Birds respond to cold air by contracting these muscles, and the

33

raising of the feathers makes the bird look fatter. Normally the feathers are waterproofed with an oily secretion and so do not become wet; but if they do, then the bird may die of cold.

21.3.4 Birds

The suggestion that the first advantage to birds of the evolution of feathers was probably in the control of body temperature may surprise you. Feathers have been exploited by birds in the construction of efficient wings, which are modified forelimbs—see Figure 19.

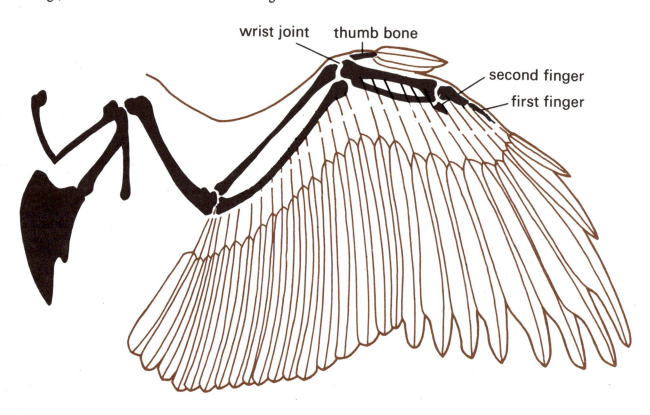

wrist joint   thumb bone

second finger

first finger

Figure 19 *Diagram to show the structure of a bird's wing.*

The earliest known bird, fully feathered but with a long, bony tail, was found in rocks about 160 million years old. It seems likely that its ancestors lived in trees and leapt from branch to branch, holding out their forelimbs. Those with long feathers on these limbs were able to leap further and so, gradually, the birds may have evolved into gliders (see section 21.4). Flapping of the wings during the glide may have enabled some birds to propel themselves through the air. The vast majority of birds today fly using muscular effort and beating the wings (see section 21.4.1).

The design of modern aeroplanes must take account of the probable use of the plane—supersonic airliners differ in shape from large cargo planes or small light aircraft. Similar aerodynamic principles apply to birds, and the different shapes of their wings can be related to different habits, i.e. the birds show considerable radiation in wing shape, as illustrated by Figure 20 (margin and opposite).

Swifts and swallows belong to different families. Both feed on insects which they catch during flight and both are highly manoeuvrable. They look very much alike.

Swift

Swallow

Hummingbirds can hover as they suck up nectar from flowers.

Gulls spend much time soaring in rising air currents over the sea or by cliffs.

Eagles can soar but also swoop and lift up prey in their claws. They are powerful fliers and can carry comparatively large prey. Owls fly almost silently and can swoop and lift up a small rodent in their claws.

Pheasants spend most of their time on the ground. They can zoom into flight when threatened by danger but they can fly for a short time only.

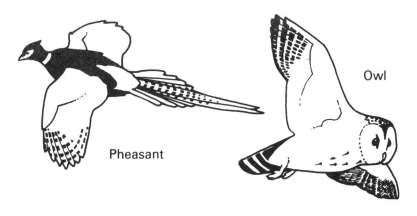

Falcons are hunters that outfly their prey and snatch it out of the air.

Penguins hunt fish under water, using their modified wings for flying through the water. They cannot fly through the air.

Figure 20   *Modern birds illustrating different shapes of wings and patterns of flight.*

There has been considerable adaptive radiation of the feet of birds. Look at Figure 21 comparing the feet of a duck, a hen, a hawk and a robin. The duck has webbed feet, used for swimming; the hen has stout feet with short claws, used for scratching the earth; the hawk has very strong hooked claws which it uses for grasping prey and also for perching; the small song birds have quite long toes and short claws and use the feet for perching.

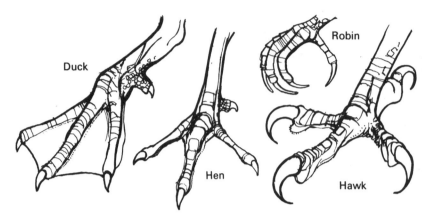

Figure 21 *Four types of birds' feet.*

Birds' beaks also vary very much in size and shape. Since the beak is the principal organ used to collect and manipulate the food, it is not surprising that the particular size and shape of beak can usually be related to the sort of food eaten. You have already met examples of this among Darwin's finches. Others are shown in Figure 22.

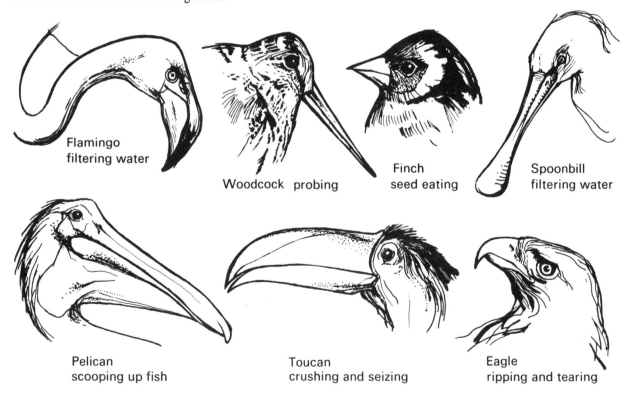

Figure 22 *Heads of birds with different feeding habits.*

Adaptation of parts of an animal's body clearly occurs—in the shapes of the wings, feet and bills of birds, for instance. While it is often convenient to discuss either wings or feet or bills as examples of adaptation, it is important to realize that these are parts of a whole organism, in this case a bird. Adaptation is a phenomenon of the whole organism, not just of one or a few of its parts.

Consider a peregrine falcon. Peregrines hunt and catch some of their prey in the air. They are fast flying birds and their hunting technique is to fly high and then, having located suitable prey below, to dive down and strike the bird prey at the back of the neck with their claws. Then, as the swooping falcon is travelling much faster than the dead body of its prey, the peregrine whirls past and turns upwards to recover its capture in the air (Fig. 23).

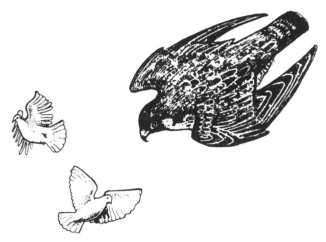

Figure 23 *A peregrine falcon 'swoops' after two pigeons.*

This style of hunting depends on the peregrine's ability to fly faster than its prey, who themselves are often fast flying birds such as pigeons. The falcon has acute eyesight—it often starts its downwards swoop while more than 60 m above the prey. The fast swoop through the air imposes strains and stresses on the feathers and wings and the delicate surfaces of the eyes are protected. Striking the prey at high speed can be successful only if the feet, legs and claws are strong enough to withstand the sudden impact. High speed aerobatics require great control of the body muscles and sensitive balance; both these depend on well-developed sense organs and brain. Finally, ripping the body of the prey into bite-sized chunks requires a strong beak and powerful neck muscles. These strong muscles imply strong bones to pull against—thus the head and neck skeleton must also be adapted. And in addition to all these, there must be appropriate adaptations of the circulatory and digestive systems—in fact, the whole body is involved.

All parts of the organism must be integrated for that organism to survive—and survival involves behavioural as well as structural and physiological adaptations. It is the whole organism that is subject to natural selection, not parts of its body; it is the whole organism that evolves.

Nevertheless, it is useful to consider the adaptation of different parts of the body separately. Each organism lives in a specialized way and in a special habitat—that is, it is adapted to a niche. This niche may overlap in some ways with other niches; each is a unique combination of relationships found in other niches. Thus an organism combines a number of adaptive features, each of which may be found in organisms of the same group filling overlapping, but different, niches. Where there are large numbers of parts of the body and physiological and behavioural characters showing adaptation, it is possible to combine these in a very large number of permutations—and thus to produce a very large number of species, each of which occupies a narrow niche. There are more than 8 000 species of birds, and their successful adaptive radiation is the result of adaptive modifications of beaks, feet, wings, physiology and behaviour.

Compare the kingfisher and the gannet. Look at filmstrip 21a: 1 and 2, and at Figure 24. Both birds feed on fish, which they catch by diving into the water. Their beak shape and skull architecture are very similar.

Gannet   Kingfisher

Figure 24   A comparison between two fish-eating birds, the gannet and the kingfisher. (See also filmstrip 21a: 1, 2.)

Compare their wings – suggest their habits.

Gannets have long narrow wings and spend much time soaring over the sea; kingfishers have short wide wings and fly by active flapping for short spells only.

Compare their feet – suggest their habits.

Gannets have webbed feet; they can land on the sea and paddle short distances and can walk on land. Kingfishers have clawed feet and perch on branches.

Their nesting habits are totally different: kingfishers nest in holes in the bank of rivers; gannets nest on cliff ledges or flat ground, making a pile of seaweed or flotsam. The kingfisher hovers or fishes with a shallow dive from a perch; gannets fold back their wings and dive into the water from about 30 m.

Here then are two species of birds, each a mosaic of adaptive characters, overlapping in a few features, but contrasted in all others.

21.3.5 Mammals

viviparity

Mammals have a number of special anatomical and physiological characteristics; two of the most significant are the development of *viviparity* and *mammary glands*. Viviparity means that the fertilized egg is retained within the body of the mother, the growing embryo being nourished by food supplied from the mother and born in a well-developed state. In mammals, the embryo develops in the uterus; food is supplied through the placenta, to which it is carried in the mother's blood and from which it is collected by the blood of the embryo. Oxygen reaches the embryo by the same route and waste products of the embryo are eliminated, via the placenta and mother's blood system, through the mother's kidney (Fig. 25).

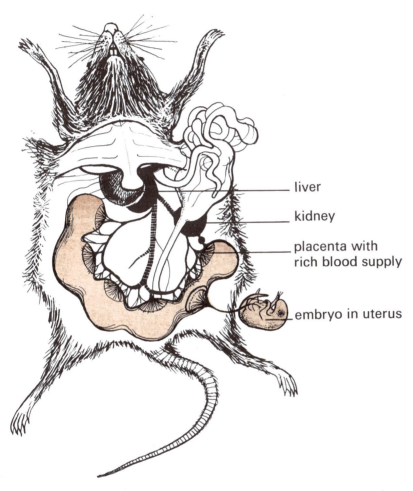

— liver

— kidney

— placenta with rich blood supply

— embryo in uterus

Figure 25   A pregnant female rat dissected to show the young developing in the uterus.

mammary glands

The young mammal feeds after birth on milk, which is secreted by special glands, the mammary glands. After a longer or shorter period when it remains wholly dependent on its mother for food, the young mammal begins to feed on the diet characteristic of its species—grass for calves and foals, meat for kittens and puppies, and so on.

Viviparity is not confined to mammals. Many sharks are viviparous, and a few have structures resembling the placenta. A few teleosts, amphibians and reptiles also are truly viviparous, but there are no viviparous birds. Only the mammals and birds supply food to the young after birth or hatching—birds supply food (usually partly digested) through their mouths; mammals have mammary glands producing milk. Some teleost fishes guard the young for a short time after hatching.

The duck-billed platypus and the spiny ant-eater are called mammals because they have fur and produce milk, but they lay eggs and incubate these and then feed the young on milk after they have hatched out. They represent a parallel evolution to other mammals descended from similar reptilian ancestors.

Fossil mammals are recognized by certain features of the teeth, jaws and ears (section 21.2.1). Some fossils with these features are 190 million years old, but the real radiation of modern placental mammals did not happen until more than 100 million years later—after the large and varied reptile fauna had become extinct (Appendix 1).

How is viviparity of survival value to mammals?

The mother must eat enough to feed the young, but she does not have to produce all the required food at once, as with the hen laying an egg. She is vulnerable while giving birth, and her activity may be hampered when the embryos are almost ready for birth and when she is suckling young. The contact between mother and young—sometimes also with the father—allows the young to learn certain types of behaviour by imitation. This is correlated with the development of certain centres in the brain. From what you read in Unit 18 about brain structure and function in man, you probably remember that the brain is organized into various centres. Some of these receive information from sense organs; some correlate information from other centres; some pass instructions to effector organs. Most of these centres occur in a part of the brain called the *cerebral cortex*.

The cerebral cortex can be recognized in the brains of reptiles and birds, but it is not very extensive in either group. Birds have a very large region of the brain called the *striatum*, and this is probably where most of their 'higher centres' are located. Birds show complex behaviour patterns, but much of their behaviour is innate, that is, it is inherited and cannot be modified to any great extent as a result of experience. But birds do exhibit some modifiable behaviour.

The behaviour of mammals includes many innate responses to external situations, but also many responses modified as a result of experience. Mammals can learn, some better than others. Their ability to learn is largely based on the cerebral cortex, and this is more extensive in some groups of mammals than in others. One group of mammals showed a progressive increase in the relative size of the brain and in the complexity of the cerebral cortex—this group is called the Primates and it includes the monkeys, apes and man.

The developing young carried in the body of the mother are protected, kept at a constant temperature, and supplied continuously with food and oxygen.

brain centres

21.3.6 The primates

The earliest primates were probably small animals living in trees and feeding on a mixed diet of insects, supplemented by fruit. They probably behaved and looked rather like squirrels, as the tree shrews do today. Their descendants also lived in trees, but developed hands, used for grasping, with nails instead of claws. They may have used the hands for conveying food to the mouth, and they show a shortening of the snout. They had large eyes; these were turned to look forwards and their fields of vision overlapped, allowing stereoscopic vision (Unit 2). This makes judgment of distance simpler, and so would be an advantage to an arboreal animal jumping from branch to branch. It is not difficult to picture how the descendants of these early primates became monkeys—living in trees,

tree shrews

monkeys

40

able to grasp branches and food (sometimes insects, sometimes fruit and leaves), and using their eyes to guide these movements. Some monkeys are quite small, such as marmosets (weight about 2 kg), but some are much larger, such as baboons (weight about 50 kg). Baboons actually live most of their lives on the ground.

One of the most obvious distinctions between monkeys and apes is the absence of a tail in the latter. Gibbon, chimpanzee, orang-utan and gorilla all have arms longer than their legs and all can swing themselves from branch to branch through forests. All can walk on the ground using their hind legs. Gibbons run holding their long arms above their heads, but the others usually walk using the arms as legs, with the fingers turned back and the knuckles on the ground. Man is typically a ground-living animal and differs from apes in that his legs are longer than his arms and he always walks upright with his arms at his sides. The upright posture is associated with many modifications of the skeleton and muscles.

The apes and man retain the five fingers and five toes and almost all the bones found in the fore and hind limbs of the earliest mammals. In this and in other anatomical features, they are much less modified from the primitive mammals than are for instance, the horse and cow (Fig. 26). The horse has one fully developed toe on each foot and the cow has two on each foot—the hooves represent nails on these digits. The retention of primitive limbs with five digits has been exploited by the monkeys and man in many niches in the trees and a few on the ground.

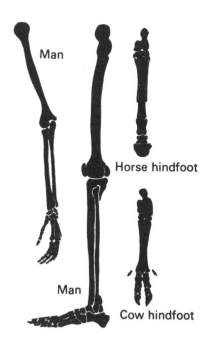

Figure 26 *The bones of the human arm and leg compared with the bones of the feet of a horse and a cow.*

The brains of the later primates are far from primitive. There has been an increase in the relative size and complexity of the brain in all living mammals when compared with their early ancestors, but the change is very much greater among the primates. It is possible that this increase in brain capacity was of special advantage to arboreal animals. The result has been that primates are able to survive in the struggle for existence by the use of behavioural responses (e.g. cunning), whereas other mammals have very specialized anatomical modifications—such as the hooves of horses and cows which allow them to gallop away from predators. Man inherited a large cerebral cortex and has evolved into an animal with much greater capacity for mental activity than any other living form. The power of speech, and the use of language to communicate information and ideas as well as emotions, result from this. In early types of man an important stage was the use of tools in hunting and feeding and later in cultivation of plants. Chimpanzees use sticks and stones in a simple way.

Some monkeys and apes have fairly elaborate social behaviour. Modern man uses elaborate tools and has complex societies (this was discussed in Unit 1). Anatomically he is distinguished from earlier types of men by his greater cranial capacity—meaning that his brain is relatively larger, especially the cerebral cortex.

Man has gained a very much greater control over the environment than any other species of animal. Modern medicine allows individuals to survive and reproduce who would earlier have been eliminated at birth or in youth. Life expectancy at birth in Britain and the USA is now twice what it was one thousand years ago. Human societies evolve and are themselves subject to natural selection. The implications of these changes in selective pressures lie in the field of social science, but we shall comment further in Units 33 and 34.

Now you can do SAQs *1 to 3.*

21.4 Convergence

So far we have been discussing the evolution from a common ancestor of descendants which differ from each other and which live in different niches; they show adaptive radiation and this implies divergence in structure and habits, both from the ancestor and from other lines of descendants. Each new species has features of anatomy and physiology which give it selective advantages in the particular niche which it occupies. You may wonder what happens when the descendants of two unrelated or distantly related ancestral forms occupy similar niches. Each of the descendant species must have features which are of special advantage in the niche, and this almost certainly means that the two resemble each other in certain adaptive features. This phenomenon is called *convergence*. It is the result of animals of diverse ancestry having evolved similar structures or physiological processes and so surviving in similar niches. Sometimes the similarities due to convergence are so striking that two species have been assumed to be related to each other and given similar names, and it has taken much investigation to unravel their real relationships.

The vertebrates evolved in water and the most successful and varied group of fishes, the teleosts, are an aquatic group, radiating into many different niches in the seas, rivers, lakes, swamps and springs. Very few have exploited the land or the air, and even these all breed in water. The terrestrial vertebrates, on the contrary, have evolved many forms convergent to the fishes and living in water, some for the whole of their lives, e.g. whales and sea snakes. But all these secondarily aquatic forms breathe air through lungs; this limits their exploitation of aquatic habitats, since they must rise to the surface at intervals. We shall show you some of these animals in this Unit's TV programme; they include remarkable examples of convergence.

In this text, we shall consider vertebrates that have taken to the air—either by the development of true flight (muscular flapping of the wings) or as gliders. As you might expect, various groups of terrestrial vertebrates have adopted one or other of these habits and, in fact, there are also gliders among the teleost fishes.

21.4.1 Gliding flight

The simplest type of flight is gliding, where there is no active propulsion (no power used) after the initial launching of the animal (or glider) into the air. If there are rising air currents, then it is possible for a man-made glider to soar and to keep up in the air for long periods of time—even to cover considerable distances. The human gliding record for distance in Great Britain is 579 km (Lasham, Hampshire, to Portnoake, near Edinburgh) on 10 May 1959. There are many vertebrate gliders: flying fish, a flying frog, the flying dragon (a lizard), the flying oppossum, several different families of flying squirrels. Some of these are shown in Figure 27.

What do they all have in common?

Some sort of membrane on either side of the body forming the 'wings'.

Figure 27 *A selection of modern vertebrates that glide, not drawn to scale.*

This membrane may be supported by finrays (the fish), or ribs (the lizard), or the fingers (the frog), or it may be stretched between the fore and hind limbs (the various mammals). The wings are held out when the animal glides, but do not flap. The frog, lizard and mammals are tree-living animals which leap from branch to branch or down to the ground. The wings enable them to get further away from the leaping-off point than they would if they jumped, and to reach the ground without damage.

They escape predators. They have a better chance of moving from tree to tree to obtain food.

Flying fish feed in the upper layers of the ocean. They swim rapidly just below the surface, then lift the head and expanded fins out of the water and taxi along using the lower part of the tail, gaining speed. They then lift out the tail and glide for up to 40 m before dropping back into the water again. In this way they escape from large predatory fish.

21.4.2 True flight

The power for gliding comes either from the action of gravity pulling the animal down to Earth or from fast movement developed in another medium (as when flying fish taxi along the water and then rise out of it and glide). In true flight, the power is developed in the animal's own muscles, which flap the wings and so generate a pattern of air flow such that the body can move smoothly forwards or rise or descend in the air. Man-made flying machines usually have fixed aerofoils, but flying animals move theirs; this type of movement is different from that of a helicopter's rotor blades, since the blades spin with a circular motion whereas wings beat up and down. As described in section 21.3.4, the size and shapes of birds' wings are adapted to the habits of the animal, and this applies to other flying animals besides the birds. Apart from vertebrates, one other group of animals have evolved true flight—and very successfully! This group is the insects; you can read about them in section 21.5. Here we shall study convergence between the flying vertebrates only.

The three groups of flying vertebrates are the birds, bats and pterosaurs. The pterosaurs are all extinct (see Appendix 1); there are no fossil remains of their soft parts, so their physiology is unknown. Read section 21.3.4 again if you need to be reminded about the habits of birds. Bats are mammals; they are furry, warm-blooded, viviparous and suckle the new-born young. British bats are all nocturnal and eat insects, which they catch on the wing, using echo-location (refer back to Unit 2); but elsewhere there are bats which feed on fruit or nectar or blood, and even bats that catch and eat fish.

Compare the adaptations to flight shown by birds, bats and pterosaurs by inspecting Figure 28 and then answering the questions below.

1 Apart from skeleton and muscles, what makes up the wing?

2 The forelimb supports the wing in all three types. Is the hindlimb also involved?

3 How many fingers support the wing?

4 Is it likely that the hindlimb will show adaptive radiation?

1 Birds—feathers; bats and pterosaurs—skin.

2 Birds—no; bats and pterosaurs—yes.

3 Birds—2 (II and III; I forms bastard wing); bats—4 (II, III, IV and V; I is clawed); pterosaurs—1 (IV; I, II, III are clawed, V absent).

4 Birds—yes (see section 21.3.4); bats and pterosaurs—no, because these limbs are part of the spreading mechanism of the wing, and adapted to that function.

The forelimbs of birds, bats and pterosaurs have a basic structure which can also be seen in the forelimbs of crocodiles, lizards, turtles and such diverse mammals as rats, whales, horses, cows, elephants and man. This basic forelimb has undergone three different types of modification in the three types of wing. The convergence between bats and pterosaurs, both with wings of stretched skin, is more marked than between these two and the birds, even though the birds and pterosaurs are more closely related to each other than either is to the bats.

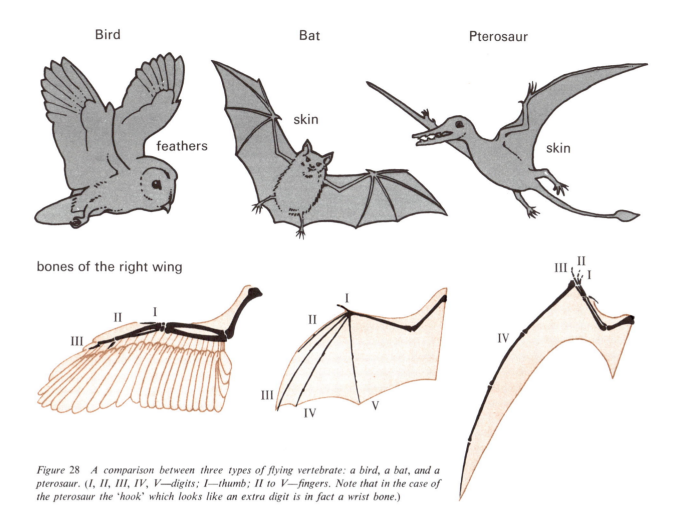

Bird

Bat

Pterosaur

bones of the right wing

Figure 28   A comparison between three types of flying vertebrate: a bird, a bat, and a pterosaur. (I, II, III, IV, V—digits; I—thumb; II to V—fingers. Note that in the case of the pterosaur the 'hook' which looks like an extra digit is in fact a wrist bone.)

Turn to black-page Appendix 2 now, if you wish to study more examples of divergence and convergence. The exercise is based on the unique mammal fauna of Australia; it is intended to give you practice in the principles underlying sections 21.1 to 21.4.

Now you can do SAQ *4*.

21.5 Arthropods and Molluscs

Many of the vertebrates are large and conspicuous animals. Being ourselves vertebrates, we are more familiar with the pattern of structural organization within this group than with the many other patterns of animals which exist today. All these must be considered to be successful types; some of them could be considered to be more successful than the vertebrates, depending on what criteria are used to assess success.

If success is judged by the number of species described for the group—and this is really a measure of the number of different niches occupied—then the insects are outstandingly successful. There are more species of insects than of all other multicellular animals added together—more than half a million species! They manage to live in almost every possible habitat on land or in freshwater but, curiously enough, there are very few marine insects. Insects are usually grouped into a larger assemblage called the *Arthropoda* (from the Greek 'jointed legs'), which also includes crustaceans, spiders, millipedes and centipedes. All these have a hard 'shell' round their bodies (an *exoskeleton*—'outside skeleton') and the legs are fitted to the body by joints and are themselves jointed. You probably have a general idea of what these animals look like. Some are shown in Figure 29 and Filmstrip 21a, 3 to 8. Notice the different numbers of legs in the different groups of arthropods. If you have started the Home Experiment for Unit 20, you probably have some specimens of the fruit-fly, *Drosophila*; you can examine these to see the jointed structure of the legs and other parts of the body. Note that the flies are the only group of insects to have only one pair of wings—most adult insects have two pairs. In Units 19 and 20 you have met several insects, notably moths, and you will remember that the egg hatches into a caterpillar which looks quite different from the adult moth. It is a characteristic of many insects that they have young stages which are very different in appearance from the adult. You should generally find it easy to recognize an adult insect as such—after all these are the only non-vertebrate (*invertebrate*) animals to have evolved flight—but you might well find difficulty in recognizing some of the larvae.

Another very successful group is the *Mollusca* (from Latin 'soft'): the snails, mussels, squids and their allies. Here the special features are a soft body (usually, but not always) protected by a shell or shells. Molluscs are a group which has principally diversified in the sea; there are relatively few species living on land (see Fig. 30 and Filmstrip 21b: 9, 10 and 11 for examples).

At the present time, insects are very abundant on land and almost absent from the sea, whereas molluscs and crustaceans are abundant in the sea and almost absent on land. This difference could be the result of the evolutionary history of the groups. For example, land-living crustaceans could have been abundant in the past and have become extinct, and there might have been marine insects in the past, now also all extinct. Study of fossils does not support this—the earliest known fossil crustaceans are marine forms; marine crustaceans were abundant and land ones very uncommon in the past, as in the present. Molluscs, also, are common as marine fossils; they date back some 500 million years. Insects, on the contrary, are typically found as fossils in land or freshwater deposits.

arthropods

exoskeleton

insects

molluscs

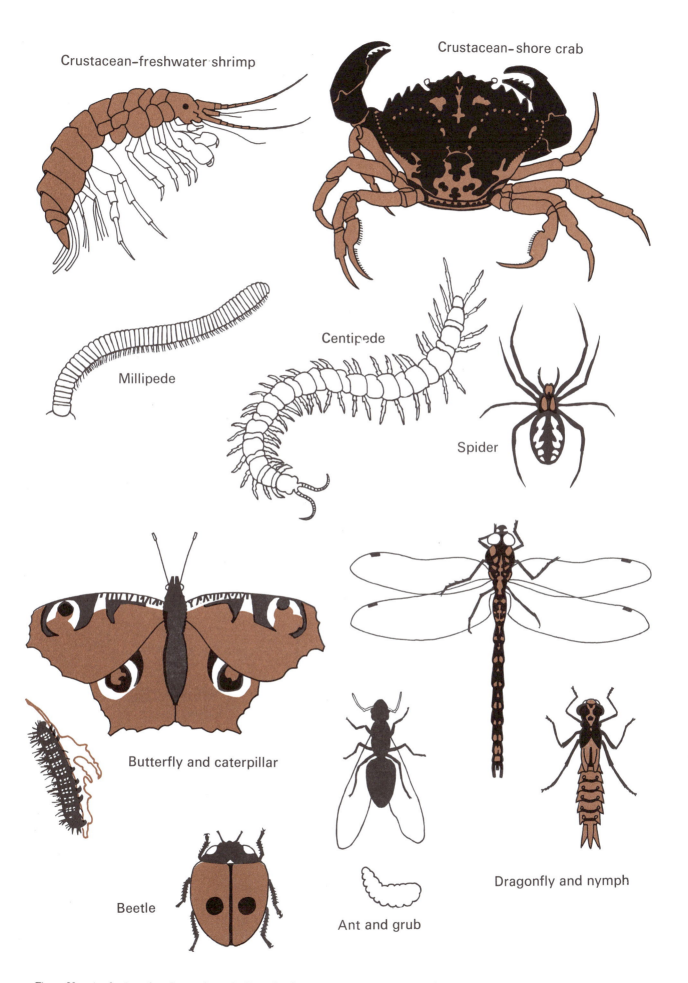

Crustacean-freshwater shrimp

Crustacean-shore crab

Millipede

Centipede

Spider

Butterfly and caterpillar

Beetle

Ant and grub

Dragonfly and nymph

Figure 29   A selection of modern arthropods. (See also filmstrip 21a: 3, 4, 5, 6, 7, 8, and 21b: 12, 13, 14.)
The lower seven are insects.

The earliest true land plants and the earliest insects are found together in rocks over 400 million years old. About 100 million years later, the coal measures were laid down during the Carboniferous Period (see Table 3, p. 23), as the fossilized remains of a lush vegetation of giant mosses, ferns and similar plants. The evolution of a land flora made available new niches for animals, and there were many diverse insects living at the time when the coal-measures' plants flourished. The swampy pools were full of bony fishes and the earliest amphibians and reptiles were beginning to colonize the land (see section 21.3 and Table 3).

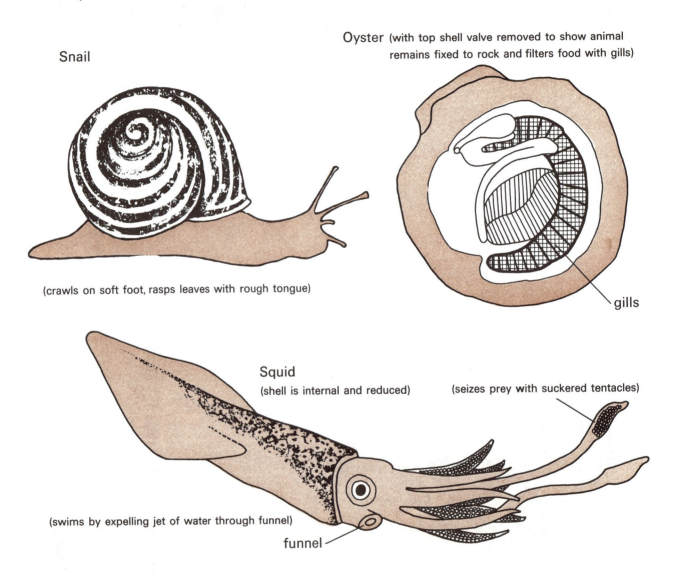

Snail

Oyster (with top shell valve removed to show animal remains fixed to rock and filters food with gills)

(crawls on soft foot, rasps leaves with rough tongue)

gills

Squid
(shell is internal and reduced)

(seizes prey with suckered tentacles)

(swims by expelling jet of water through funnel)

funnel

Figure 30   *Three modern molluscs. (See also filmstrip 21b: 9, 10, 11.)*

Comparing insects with molluscs again, the insects can be seen to have attributes which seem very suitable for terrestrial life. Insects have a light exoskeleton covering the whole body and suitable for water-proofing; extensions of this exoskeleton form the wings. The soft-bodied molluscs, when living in water, can move in spite of the heavy shell; but such movement is much more of an effort on land, since air is a less dense medium than water. So the protective shell may become a liability instead of an asset. Insects have a system of tubes conveying gaseous oxygen to all the cells of the body and this, together with the design of their jointed limbs, imposes a maximum size limit. The largest living species are beetles about 160 mm long (about three times the size of the *smallest* living mammal, the masked shrew). Some of the extinct insects found in the coal forests were much larger, with wing spans up to 700 mm. Contrast these with the largest living molluscs, which are squids with bodies about 6 m long, and arms adding a further 12 m to their total length. These can be compared with fishes and whales—the whale shark may reach lengths of 15 m and weigh several tonnes, while the blue whale grows much larger, reaching 30 m in length.

There was a further evolutionary radiation of insects during the Mesozoic Era, when the reptiles were dominant on land. The earliest flowering plants are found towards the end of this Era (Table 3, p. 23). The variety of flowering plants provided a variety of new food sources and feeding niches for animals; this was exploited by insects as well as by the mammals of the Cenozoic Era. Butterflies and moths are not found in deposits more than 70 million years old, and we may surmise that their evolution has been closely tied to that of flowering plants.

21.5.1   Co-adaptation of insects and flowers

The radiation of flowering plants is closely bound up with the radiation of insects. Before seeds are formed, pollen* grains must be transferred from the stamens* to the female part of the flower, the stigma*, from which they grow to the ovule* (which grows into the seed).

Wild Rose–produces much pollen, eaten by beetles

(a)

From what you read in Unit 19, is it an advantage for the pollen to come from the same plant as the stigma, or from a different plant?

Genetic diversity would be reduced if the pollen always came from the same plant, so it is likely to be an advantage if there is 'cross-pollination' with pollen from a different plant.

In conifers, pollen is transferred by the wind; this also happens in many flowering plants such as the grasses. (If you suffer from hay-fever you will be uncomfortably aware of this.) But many flowers are pollinated by insects—this is only possible because of the remarkable co-adaptation between these two sorts of organism.

Pollen is valuable food for insects—this sort of pollen has a sticky coat and probably a scent which is attractive to some insects. There is an over-production of pollen in all flowers, but some, such as wild roses (Fig. 31a) and poppies, produce very large quantities. These are visited by beetles, which become coated with the sticky pollen as they eat it; when they fly on to another flower of the same species, some of this pollen is rubbed off on to the stigma, and so the ovules are fertilized.

Honeysuckle–long tube with nectaries at base accessible only to some moths and bees

(b)

Many flowers have organs (*nectaries*) that secrete sugars as 'nectar', and this supplies a food source for insects. In these flowers, the stamens are arranged so that an insect visiting the nectaries becomes powdered with pollen; when it visits another flower of the same sort, the pollen is rubbed off on to the stigma. In some of these 'nectar flowers', the nectar is available only to insects with long 'tongues'—the garden columbine is such a flower, with the nectaries at the end of the spurs on the petals, and there are many tubular flowers such as those of pinks and tobacco plants and honeysuckle (Fig. 31b), with the nectaries at the bottom of the tube and the stamens and stigma further up it. The early insects had biting mouth parts, as the beetles and grasshoppers have today; tongues used for sucking up nectar have been evolved in butterflies and moths and also in bees, by different modifications of the biting type. See Filmstrip 21b: 12 and 13, which show these sucking mouth parts in use.

Figure 31   *Two wild flowers that are pollinated by insects.*

---

* *Pollen* is produced by the male parts of flowers, the *stamens*, which are usually yellow. The *stigma* is one of the female parts of a flower. When a pollen grain reaches a stigma, it grows through it and makes contact with the *ovule*. Fusion of nuclei then occurs to produce a zygote from which a fertile seed can develop.

You may wonder how insects that feed on nectar identify the flowers bearing this sugar solution. Some probably are attracted by specific scents, but commonly the insects use sight to recognize flower species. Bees and butterflies have large compound eyes and excellent vision for shapes of objects. They also have well-developed colour vision (see Unit 2). Their range of sensitivity is often not identical with that of man, but extends into the shorter wavelengths of the ultra-violet, while not including the longer red waves. Some of the red flowers that are visited by bees or butterflies actually reflect ultra-violet as well as red, and so are visible as coloured objects to both man and insects; but each reacts to different ends of the visible spectrum. The evolution of insect senses has proceeded in parallel with the evolution of flowers, from the greens of the unscented wind-pollinated forms to the yellows, blues, purples and ultra-violets of garden flowers and many sweet-scented summer 'weeds'.

You may also wonder how 'self-pollination' is avoided in flowers which contain both stamens and stigmas. Usually these produce pollen from their stamens before the stigma is ready to receive pollen; by the time the stigma is in a receptive state, the stamens have finished producing pollen, so that only pollen from other flowers is available. The nectaries are active throughout the whole period, so the flowers are visited by insects all the time.

The design of the flower is commonly adapted to the characters of the chief pollinator on which the flower now depends; there are many fantastic examples of this, but here we shall deal only with two British examples, both involving bees. 'Snap-dragons', and other flowers of similar design such as toadflax and deadnettles (Filmstrip 21b: 14 and 16), are much visited by bumble bees. When these alight on the lower lip, and push their way towards the nectaries, the stamens are bent down and pollen is smeared on the back of the bee; the young stigma remains curved well out of the way. In an older flower, the stamens have finished producing pollen, but the older stigma hangs down so that the bee brushes against it as it goes for the nectar. The snapdragon design means that small, light insects cannot get to the nectar, since only a heavy insect can open the way into the flower. The more open deadnettle design can be visited also by other insects, but only the bees are heavy enough to move the stamens. The Bee Orchid flower (Filmstrip 21b: 15) has, to us, a bizarre shape and meaningless colour, but in fact the lip mimics the form and colour of the female of certain bees—different bee orchids mimic different species of bee. The males of these species of bee are deceived; if one lands on the flower, behaving as though it were a nubile female and attempting to copulate with it, sticky bags of pollen become pressed against its head and adhere there. Should the male bee be deceived again and attempt to copulate with another, older orchid flower, then the pollen is transferred to the stigma.

Bee Orchid

While such an example illustrates the exquisite perfection of the inter-relationship between particular species of flowering plant and insect, it is the general relationship between plants and insects that is really important. The evolutionary progress of the two groups has proceeded in parallel, each group offering opportunities to the other and being in turn exploited by the other group.

Now you can do SAQs *5 and 6.*

21.6 Fossils and Evolution

Our most useful source of information about life in the distant past is the fossil record. By examining fossils, we can learn about the shapes, sizes, diets and possible habits of the organisms which died at that time; from the rocks, we can deduce the conditions and climate when the particular deposits were laid down. But it is important to realize some of the snags and special features of fossil evidence.

If the unit of evolution is an interbreeding population—a species (see Unit 19)—we can test with living organisms whether a given population is a true species or not by observing whether individuals will interbreed with those from other populations. A palaeontologist (expert on fossils) cannot apply this test to his specimens. He has no way of finding out whether fossil A, which resembles fossil B in some characters, but differs from it in others, is a member of a different species or merely a variety within a single species. Consequently, *palaeospecies* (based on fossils) are of a different type from *biospecies* (based on living organisms). They have much in common with the *morphospecies* set up by an expert working in a museum with dead specimens only; but probably there are more specimens available of the modern, dead organism than of a given fossil, and there is always the possibility of working on live specimens in the field to check diagnoses.

palaeospecies
biospecies
morphospecies

A second problem arises from the nature of the fossilization process. To become a fossil, an organism must die in the right place and under the right conditions. Since this may be described as a rare event, it seems probable that the organisms that we study as fossils were common organisms rather than rare ones—the chance of a rare event happening to an uncommon animal must be much less than that of a rare event happening to a common animal. But the palaeontologist simply cannot tell whether his fossil was rare or common when it was alive. Sometimes, of course, he finds large numbers of one species in one place, and sometimes one species is found in many different places, suggesting that it was common at that time.

So the fossil record is patchy and its interpretation is difficult. This does not mean that palaeontologists are wasting their time, but that, in inexpert hands, fossil stories can become misleading. The family tree type of diagram (Fig. 32a), purporting to show the exact steps in the evolution of any organism, should be treated with reservations. It imples much more than can, from the nature of fossils, ever be known for certain. At the species level, it is more helpful to think of 'family clouds' (Fig. 32b). This is a much less definite representation, putting into diagram form a statement such as 'that organism A evolved from unknown ancestors which were possibly more like B than either C or D—but we do not know how representative either A or B were of their populations'.

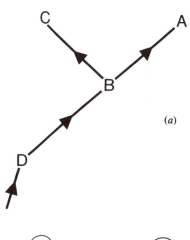

(a)

At the level of major groups, also, the evolutionary 'trees' postulated by Darwin and others must be rejected. They pictured the first organisms giving rise to progressively 'higher' types of plants and animals, with man at the top of the animal tree and flowering plants at the top of the plant tree. Look back to Figure 11 and also at Figure 33 (Appendix 1, p. 57), which show the distribution in time of fossil vertebrates of various categories: each diagram is more like a shrub than a tree. The implication of this is that the evolution of a major innovation—adaptive change—is

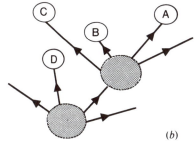

(b)

Figure 32 Two different ways of expressing evolution using diagrams: (a) a family tree (b) a family cloud.

followed by adaptive radiation as the organisms, as it were, exploit their genetic advantages. One of the radiating 'lines' may undergo the next major adaptive change, and so there will be another radiation of types.

When we attempt to relate major groups such as the vertebrates, arthropods and molluscs, we have to postulate a very simple animal as common ancestor—again, the evolutionary lines must have branched like a shrub, not as in a tree. Fossils of all the major living groups of invertebrates can be found in rocks 570 million years old. Before that, the record is more difficult to read and we can only speculate about the really distant past.

The fossil records show that, as the Earth has aged, so the diversity of living organisms has increased. If we assume, as Einstein put it, that 'God does not cheat'—that the rules of evolution and regularities of natural selection we can discover now are themselves unchanging, then, if we could go back far enough in time, we should find simpler and simpler living organisms. Ultimately, we should reach a point in time when nowhere on Earth were there objects sufficiently well organized to be recognized as living organisms.

21.7 Biochemical Evolution

So far we have considered the process of evolution operating at the level of organs and organ systems, so that more complex, diverse or 'better fitted' organisms have evolved from simple ones. You may wonder: how far have their biochemical processes and internal cell structure evolved too?

In Units 14 to 16, we described something of the hierarchical behaviour and structure of individual cells, and we did so as if these cells were somehow anonymous—they belonged to no particular tissue or organism. We discussed the structure of 'typical' plant and animal cells, but it is fair to ask how far such cells and their behaviour are indeed typical of the widely different living forms that have evolved. To answer this question, turn again to *The Chemistry of Life*, and read pp. 237–244. Much of this material will already be familiar to you in a different context, particularly the discussion of photosynthesis, which you met in Unit 15. You read about autotrophes and heterotrophes (as producers and consumers) in Unit 20.

In the TV programme for Unit 18, you saw organisms of small size that are not divided into cells. Look again at Filmstrips 18a and 18b if you need to remind yourself about what they look like. Some of them are green, like *Euglena*, and normally perform photosynthesis, while others are colourless, like *Paramoecium*, and engulf other organisms. Clearly, *Euglena* is an autotrophe and *Paramoecium* is a heterotrophe—the arche- **protistans** typal plant and archetypal animal, producer and consumer. Yet, if *Euglena* is kept in the dark, but provided with acetic acid, a simple organic compound, it can survive and grow. It loses its chlorophyll and lives as a heterotrophe. If brought back into the light, it may continue to grow and multiply as a colourless heterotrophe, swimming with its cilium. There are many other unicellular organisms that can function as autotrophes or heterotrophes according to environmental conditions, so that there is no point in trying to classify them as either plants or animals.

Euglena contains within its body the organelles that are typical of cells in general, such as nucleus, mitochondria and endoplasmic reticulum. It also has chloroplasts similar to those of large plants. Bacteria are very **bacteria** much smaller—a single bacterial cell is about the same size as a single chloroplast of *Euglena*. This makes it difficult to study them with a light microscope, but observation under the electron microscope shows that bacteria differ from larger cells in lacking internal membranes. This means that they have no distinct nucleus or distinct mitochondria.

A few bacteria perform photosynthesis and their chloroplasts lack membranes. In some of them the substrate is not water but hydrogen sulphide (H_2S instead of H_2O). Most bacteria are decomposers and they show an enormous range of synthetic abilities; some can fix atmospheric nitrogen whereas others are 'denitrifiers', releasing gaseous nitrogen from nitrates and nitrites. It is interesting that there is such a diversity of metabolic pathways among organisms that are structurally very simple.

We have emphasized the remarkable similarity of the basic biochemistry **similarity of the basic biochemistry** of living organisms. When one thinks about it more deeply, however, such similarity should not be surprising; in fact it should be seen as essential to life. Think, for instance, what would happen if the protein of one species were composed of d-amino acids and another of l-amino acids, or the carbohydrates of one species of d-sugars and another of l-sugars. Each would lack the enzymic apparatus to manipulate the chemicals of the other; each would be poisonous to the other. If plants make d-glucose during photosynthesis, then animals which eat the plants are committed to utilize the d- and not the l-glucose. The interdependence of life, through food chemicals, imposes a unity upon biochemistry.

However, despite the overall similarities, there are also a number of very interesting biochemical *differences* between species: differences which can indeed be studied, at the cellular level, by following the processes of evolution which have led to them.

biochemical differences

Cytochrome c has been studied in many different species by now. Of course, it occurs practically universally—in animal and plant mitochondria and even in aerobic bacteria (though they have no discrete mitochondria). The protein sequence for the cytochrome c of any one species is always the same. You will recall that the protein sequence is determined by the sequence of bases on RNA, and RNA is in its turn determined by DNA in the chromosome.

What is implied, at the DNA level, by a change of one amino acid in a protein sequence?

A change in one base pair in the DNA—i.e. a mutation.

You may now turn to black-page Appendix 3, which is an exercise in which you use information about the sequence of amino acids in part of cytochrome c of seven organisms. From the genetic code, described in Unit 17, you can work out the minimum number of mutations necessary to transform each of the seven sequences into that found in human cytochrome c. You can then rank the organisms in order of increasing number of mutations required, and compare this order with one based on information about evolution given in sections 21.2 to 21.5.

The conclusion from the exercise is that species that are more closely related have sequences of amino acids in their cytochrome c that are more similar than those of species which are more distantly related.

A simple mutation will change one base in a DNA triplet and so probably will change one amino acid in a protein. Obviously, if this amino acid is functionally necessary—if it is at the active centre of an enzyme, for instance—the mutation will result in an impaired organism, and may be lethal.

Are the amino acids near the N-terminal end of cytochrome c necessary or not necessary?

Presumably not necessary, since they show differences in the organisms listed.

Evolution and transformation of species must generally involve a very large number of such individual mutations.

Now you can do SAQs 7 to 9.

21.8   The Origin of Life

When we try to account for the origin of living organisms in a non-living world, we have two types of explanation open to us. First, that the evolutionary processes we have followed backwards in time can be projected still further back—that life evolved from non-living material. Secondly, it is possible to maintain that some supra-natural agency started the whole process and created life from non-living materials.

The great disadvantage of the second hypothesis is that it cannot be tested, and consequently it is useless, scientifically speaking. The first hypothesis can be tested in principle: if we can recreate the conditions that existed on Earth more than 600 million years ago, then it might be possible to find out whether living organisms could develop from non-living things. Failure to make living from non-living does not prove that such a process did not happen. Our guess as to just what the conditions were like on Earth long ago may be wildly wrong, or even not quite right, and that could prevent the experiment working.

Now read The Chemistry of Life, *pp. 247 to 252, starting at the top of page 247.*

In the TV programme, you will see the sort of apparatus used by Miller in his experiments. You have already carried out a separation of amino acids in the Home Experiment for Unit 13, so you have experience of the sort of methods by which Miller identified the compounds synthesized from the methane-ammonia mixture in his apparatus.

As part of your Home Experiment for this Unit, you will observe the formation of coacervate drops. You should not think that the coacervate drop sequence described in *The Chemistry of Life* is the *only* possible explanation for the origin of life, for there have been other theories— some based, for instance, on a 'dry' (volcanic) origin rather than a 'wet' one. You can read more about these in the Earth Sciences Reader, *Understanding the Earth.* The intention here is to present one *possible* theory for the origin of life. Nor should one underrate the difficulties in the way of providing a full account of the evolution of so complex a system as the genetic code. The picture of this 'inevitable' emergence of life that we have painted here raises a variety of interesting philosophical questions. If this process is inevitable, could it have occurred on other planets than Earth; if so, what sort of life forms might we possibly expect? Could they be intelligent? It is interesting that much of the modern research on the origin of life being conducted in the USA is financed by the American Space administration for what it might reveal of 'exobiology' (as it has been called).

coacervate drops

Some of these questions will be raised again in succeeding Units, where you will study aspects of earth and solar-system history in more detail.

Now you can do SAQs *10 to 13.*

Summary

Study of finches on the Galapagos Islands leads to the conclusion that the thirteen species found now are descendants of one or a few South American finch species that reached these isolated islands and underwent adaptive radiation (21.1).

Modern mammals include a variety of species with different habits (21.1); but they share many basic anatomical and physiological similarities implying a common ancestry. Study of fossil mammals reveals that early forms were different from those alive today (21.2.1).

The evolution of the vertebrates is comparatively well documented from the fossil record; it can be interpreted as the appearance of a series of innovations followed by adaptive radiation of the groups possessing these new features (21.3). Among modern vertebrates, teleost fishes, birds and mammals show considerable diversity of form and habit. The varied reptiles of the Mesozoic Era are described in Appendix 1.

One consequence of adaptive radiation is that unrelated animals may occupy similar niches and show remarkable convergence; this is illustrated by the flying vertebrates, the birds, bats and pterosaurs (21.4). In Appendix 2, convergence and divergence are both illustrated by comparing the marsupial mammals of Australia with other mammals.

The two largest groups of modern animals, the arthropods and molluscs, are very different in basic structure from each other and from the vertebrates (21.5). Some flowering plants show remarkable co-adaptation to insects, especially to bees and butterflies (21.5.1).

There are certain drawbacks to the fossil record, but study of fossils and of modern organisms reveals that evolutionary lines probably branch like shrubs, rather than forming trees (21.6).

Arguing backwards to single-celled organisms, and considering cell structure and biochemistry, we note remarkable similarities but also some interesting biochemical differences (21.7 and Appendix 3).

Speculation about the origin of life (21.8) ends this Unit, but forms a link with the Units on Earth Sciences that will follow.

Appendix 1 (Black)

Fossil Reptiles of the Mesozoic Era

Fossils found in rocks aged between approximately 225 million and 65 million years are described as belonging to the *Mesozoic Era*. This Era is a time of special interest to students of evolution, since the teleost fishes, the mammals and the birds all evolved during this Era. But the fauna on land was dominated by reptiles, a group which radiated in a remarkable way at this time (Fig. 33). Some were extremely large, so the Mesozoic Era is sometimes called 'the Age of Reptiles'. Most of these varied reptilian types became extinct by the end of the Era; but some of the fossils are well preserved, so that it is possible to make many deductions about how the animals lived and what they looked like.

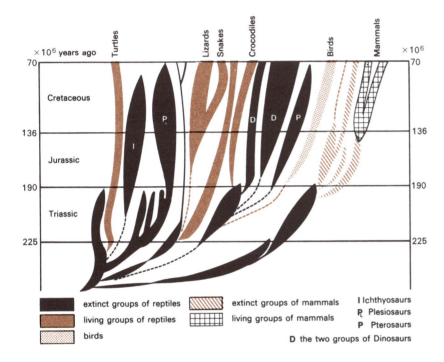

Figure 33   *Diagram to show the distribution of fossil reptiles, birds and mammals in Mesozoic rocks. The width of each part indicates the number of fossils of that group found in rocks of that age.*

There were several groups of fully marine reptiles of large size. We shall show you some of these fossils in the TV programme for this Unit.

Most of the Ichthyosaurs (Fig. 34) seem to have been fish-eaters about the size of modern dolphins, but a few fossils are known more than 10 m long. Many Plesiosaurs (Fig. 35) were larger than this, with a maximum length of about 17 m; presumably they paddled themselves through the water, whereas the Ichthyosaurs swam like fishes.

The terrestrial reptiles of the Mesozoic included the flying pterosaurs ('winged lizards') (Fig. 36). The largest had a wing span of about 8 m, but some were as small as sparrows. From their skull structure, we deduce that they ate fish, which they probably scooped up from just under the surface while gliding over shallow seas. We discuss their wings in section 21.4.2.

Figure 34   *An Ichthyosaur (Greek: fishlizard).*

Figure 35   *A Plesiosaur (Greek: near lizard).*

57

The large terrestrial Mesozoic reptiles are known collectively as dinosaurs (from the Greek: 'terrible lizard'), but anatomical studies have shown that there are two main lines. The common ancestors to these were fairly small (up to 1 m long) and looked like modern lizards, but probably stood up on their hind legs, balanced by the powerful tail. Their front legs tended to be rather short. Experts argue about whether they were amphibious or fully terrestrial in habit. Probably they were the ancestors of crocodiles and birds as well as of pterosaurs and dinosaurs (Fig. 37).

One of the lines of dinosaurs included some enormous animals—in fact, the largest terrestrial animals that have ever lived. *Brontosaurus* and its relatives were quadrupedal herbivorous forms; they may have been semi-aquatic in habit, living rather like the hippopotamus today and laying their eggs on land. One slender species was nearly 30 m from nose to tail, but probably weighed only about 10 tonnes; a shorter but more heavily built species may have attained a weight of nearly 80 tonnes. Related to these massive quadrupeds were a group of bipedal carnivorous dinosaurs. Many of these were small or medium-sized, but one of the latest of this group, *Tyrannosaurus*, was 15 m long and probably weighed nearly 7 tonnes—as far as we know, this was the largest carnivore that has ever lived on land.

Probably all the members of the second line of dinosaurs were herbivores. They did not reach the impressive size of some members of the first line, but they included some bizarre forms armed with knobs on the head or set about with protective plates or frills of bone. These armoured dinosaurs were quadrupeds. *Iguanodon*, about 8 m long, is the best known bipedal form and seems to have lived in herds—seventeen specimens were found together in one Belgian site. Another group had crested skulls, duck-like bills and webbed feet, but their stomach contents were bits of land plants—perhaps they dived into water when pursued by the carnivorous dinosaurs!

Dinosaurs and pterosaurs of varied types are found through the second half of the Mesozoic Era. Some species seem to disappear from the fossil record and others appear. At the end of the Era, they all disappeared; but lizards, snakes, crocodiles and turtles survived. Mammals and birds had already been present for more than 100 million years—and members of these two groups survived. It is still a mystery why the dinosaurs became extinct. They were widely distributed in the world, and some types had lived with comparatively small changes in form for more than 100 million years. All sorts of explanations have been suggested, but probably the most plausible are a combination of: change in climate with cooling affecting dinosaurs; change in vegetation, from ferns to grasses and herbs, affecting the herbivorous forms directly and so leading to starvation of the carnivores; increase in cosmic radiation (but why should the dinosaurs be more affected than birds and mammals?). It is interesting that there was a change in the marine fauna at the same time. This did not affect the teleost fishes, which continued to diversify and flourish in the following period.

Your Earth Science Reader, *Understanding the Earth*, includes an article about extinctions of fossil groups of animals.

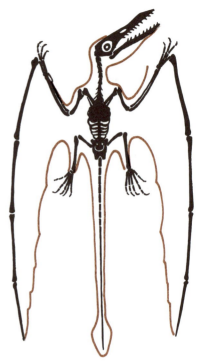

Figure 36 *A Pterosaur.* (*See also Figure* 28).

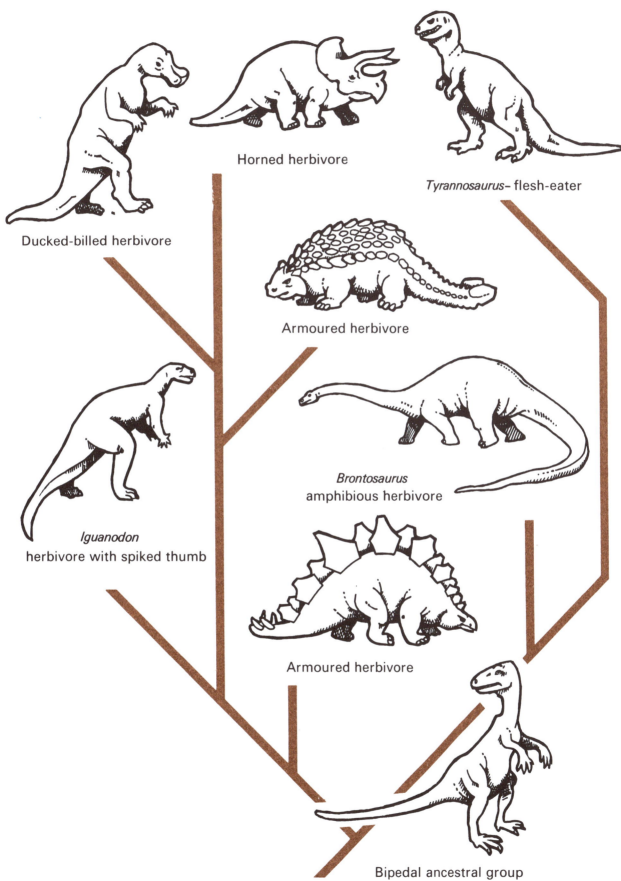

Horned herbivore

Tyrannosaurus– flesh-eater

Ducked-billed herbivore

Armoured herbivore

Iguanodon
herbivore with spiked thumb

Brontosaurus
amphibious herbivore

Armoured herbivore

Bipedal ancestral group

Figure 37   *Diagram to show possible evolutionary lines of the Dinosaurs.*

The Marsupial Mammals of Australia

An exercise to illustrate the principles of divergence and convergence.

Most of the native mammals of Australia are of the type called *marsupial* (from the Latin for pouch). In most of these animals, the female has a pouch on the belly in which the young feed and develop after birth; the young are born in a very undeveloped condition compared with the majority of mammals, the placental mammals. There are certain other differences, especially in the skeleton, between the two groups of mammals; but they are so similar that they almost certainly shared a common ancestor, probably about 100 million years ago. Fossils recognizably belonging to both of the two groups occur in rocks about 70 to 80 million years old.

Look at the 'selected marsupials' in Figure 38 (p. 62).

What phenomenon do these illustrate?

Adaptive radiation (divergence).

Look at the 'selected placental mammals' in Figure 39. Match these with the marsupials, scoring the match as 'good' or 'fair', according to whether you think the niches occupied are very similar or fairly similar.

You should get:
A	— 3	fair;	B	— 1	good;
C	— 4	fair;	D	— 8	good;
E	— 2	good;	F	— 5	fair;
G	— 6	good;	H	— 7	good.

in matching these pairs of animals—and you can also match the flying phalangers with the flying squirrels (see 21.4.1, p. 43).

What phenomenon is illustrated?

Convergence.

It is possible to produce other 'matches' for the selected placentals. For instance, besides the European mole there is the golden mole of South Africa, and there are at least two different types of 'mole rat' in tropical Africa—all these dig with forefeet with strong claws, and are covered with silky, short fur; all have eyes reduced or absent.

There are many small carnivores like the genet, and a variety of rodents like the capybara (for instance the guinea pig) and similar mammals such as conies and pikas. These additional matches usually live in different parts of the world.

How can we explain this?

The animals occupy similar or identical niches; if they lived in the same place, they would come into competition with each other and probably one species only would survive.

Consider the following items of information.

(a) The only native placental mammals of Australia are bats and some peculiar mice.

(b) Apart from Australia and nearby islands, marsupial mammals are found only in America—the opossum has a wide distribution and there are a few rat-like forms in South America.

(c) Marsupial fossils are widespread in many parts of the world in rocks of 70 to 80 million years ago; the earliest marsupials are probably those in North American rocks of 85 million years ago.

Suggest a hypothesis to account for the wealth of marsupial types in Australia today.

It appears that marsupials in Australia had very little competition from placental mammals in recent times before the arrival of man. Perhaps this has been the situation since the marsupials arrived in Australia—in the absence of placental mammals, the early marsupial colonists were able to undergo adaptive radiation, producing the varied types alive today. In the rest of the world, marsupials encountered placentals and the latter were more successful, so that marsupials became extinct in most places. So the adaptive radiation of marsupials in Australia can be explained in the same way as the speciation of Darwin's finches on the Galapagos Islands. But the Galapagos are volcanic islands separated by 600 miles of sea from South America.

Is there any real similarity between them and Australia?

Australia is a large island, separated from all other countries by sea.

Australia probably separated from other land masses more than 80 million years ago, and has been isolated ever since. So it is possible that marsupials might have arrived there before the separation, and placentals may have failed to arrive by that time.

You may wonder what happened to the native fauna when man and his domestic animals arrived in Australia. The marsupial wolf and some others have probably become extinct, and many of the surviving species are in danger.

Figure 38 *Eight Australian marsupial mammals (not to scale).*

Kangaroo (1) *is a vegetarian with a complex stomach; Koala* (2) *feeds only on leaves of some Eucalyptus spp.; Cuscus* (3) *feeds on leaves of many species of tree; Numbat* (4) *lives on the ground and feeds on termites (white ants); Tiger-cat* (5) *is an arboreal carnivore; Wombat* (6) *is a nocturnal herbivore that lives in burrows; Mole* (7) *burrows and feeds on worms and insects; Tasmanian wolf (perhaps extinct)* (8) *was a nocturnal hunter.*

Figure 39   Eight placental mammals for comparison with the marsupials shown in figure 38 (not to scale).

Bushbaby (A) feeds on fruit and insects by night; Impala (B) is a vegetarian with a complex stomach; Aardvark (C) lives in burrows and feeds by night on termites; Wolf (D) is a nocturnal hunter; Sloth (E) feeds only on the leaves of certain trees; Genet (F) is a carnivore living in trees; Capybara (G) is a nocturnal herbivore; Mole (H) lives in burrows and feeds on worms and insects.

Comparison of Cytochrome c from Man and Seven Other Organisms

An exercise using the genetic code (from Unit 17).

A sequence of 13 amino acids, near the N-terminal end of cytochrome c isolated from human tissue, is as follows:

gly–asp–val–glu–lys–gly–lys–lys–ileu–phe–ileu–met–lys

Look at the following sequences, derived from the identical region of the peptide chain of the cytochrome c of a number of other organisms:

(a) gly–asp–val–ala–lys–gly–lys–lys–thr–phe–val–glu–lys

(b) gly–asn–pro–asp–ala–gly–ala–lys–ileu–phe–lys–thr–lys

(c) gly–asp–val–glu–lys–gly–lys–lys–ileu–phe–ileu–met–lys

(d) gly–asp–ileu–glu–lys–gly–lys–lys–ileu–phe–val–glu–lys

(e) gly – asn – ala – glu – asp – gly – lys – lys – ileu – phe – val – glu – arg

(f) gly–asp–val–glu–lys–gly–lys–lys–ileu–phe–val–glu–lys

(g) gly–ser–ala–lys–lys–gly–ala–thr–leu–phe–lys–thr–arg

For each of the sequences (a) to (g), calculate the total number of mutations that are essential if the sequence is to be changed into that found in man.

For the purpose of this exercise, one mutation should be counted for each alteration in a base pair of the triplet of bases of the DNA code for amino acids. The complete DNA code, which you met in Unit 17, is given again here.

The genetic code

First letter	U	C	Second letter A	G	Third letter
	Phen	Ser	Tyr	Cyst	U
	,,	,,	,,	,,	C
U	Leu	,,	—	—	A
	,,	,,	—	Trypt	G
	Leu	Pro	Hist	Arg	U
	,,	,,	,,	,,	C
C	,,	,,	Glu	,,	A
	,,	,,	,,	,,	G
	Ileu	Thr	Asn	Ser	U
	,,	,,	,,	,,	C
A	,,	,,	Lys	Arg	A
	Met	,,	,,	,,	G
	Val	Ala	Asp	Gly	U
	,,	,,	,,	,,	C
G	,,	,,	Glu	,,	A
	,,	,,	,,	,,	G

Examples

For the change   *Ser*(ine)   into   *Pro*(line)

	AGC		
code	AGU	code	CCU
triplets	UCC	triplets	CCC
	UCA		CCA
	UCG		CCG
	UCU		

only one change is necessary: the first letter U or A must change to C. So the number of mutations required is 1.

For the change   *Ala*(nine)   into   *Lys*(ine)

	GCU		
code	GCC	code	AAA
triplets	GCA	triplets	AAG
	GCG		

two changes are necessary: the first letter G must change to A;

the second letter C must change to A.

So the minimum number of mutations required is 2.

When you have calculated the numbers of mutations required for each of the sequences, list the sequences in order of increasing numbers of mutations.

Now look at the following list of organisms:

A — silk moth; B — tuna fish; C — chicken; D — rhesus monkey; E — yeast; F — wheat; G — horse.

Rank these organisms in order of 'evolutionary distance' from man. (For instance the rhesus monkey (D) is 'closer' to man than is wheat (F).)

Now compare your two lists—the list of increasing numbers of mutations and the list of organisms in order of evolutionary distance from man.

Look at (4) of the solution to this problem (p. 67) for the identity of the organisms from which the cytochrome was obtained.

Which of the following possible conclusions from these data is supported by your calculations?

(I) Species that are more closely related have sequences of amino acids in their cytochrome c that are more similar than those of species that are more distantly related.

(II) There is no obvious correlation between evolutionary distance from man and the similarity between amino-acid sequences in cytochrome c.

Turn back to section 21.7 of the Unit for a final comment on this problem.

Solution to the cytochrome problem

1 Comparison of each sequence with that of man. The differences between amino acids occupying the same positions in each sequence are:

   (a) glu/ala; ileu/thr; ileu/val; met/glu.

   (b) asp/asn; val/pro; glu/asp; lys/ala; lys/ala; ileu/lys; met/thr.

   (c) none—chains are identical—hence no mutations required.

   (d) val/ileu; ileu/val; met/glu.

   (e) asp/asn; val/ala; lys/asp; ileu/val; met/glu; lys/arg.

   (f) ileu/val; met/glu.

   (g) asp/ser; val/ala; glu/lys; lys/ala; lys/thr; ileu/leu; ileu/lys; met/thr; lys/arg.

2 Comparison of triplets for each pair of amino acids.

glu	ala	ileu	thr	ileu	val	met	glu
GAA	GCU	AUU	ACU	AUU	GUU	AUG	GAA
GAG	GCC	AUC	ACC	AUC	GUC		GAG
CAA	GCA	AUA	ACA	AUA	GUA		CAA
CAG	GCG		ACG		GUG		CAG

1 change, A/C	1 change, U/C	1 change, A/G	2 changes, A/G, U/A

val	pro	glu	asp	lys	ala	ileu	lys	asn	asp
GUU	CCU	GAA		AAA	GCU	AUU	AAA	AAU	GAU
GUC	CCC	GAG		AAG	GCC	AUC	AAG	AAC	GAC
GUA	CCA	CAA	GAU		GCA	AUA			
GUG	CCG	CAG	GAC		GCG				

2 changes, G/C U/C	1 change AorG/UorC	2 changes, A/G, A/C	1 change, U/A	1 change A/G

met	thr	val	ala	lys	asp	lys	arg
AUG	ACU	GUU	GCU	AAA	GAU	AAA	AGA
	ACC	GUC	GCC	AAG	GAC	AAG	AGG
	ACA	GUA	GCA				CGU
	ACG	GUG	GCG				CGC
							CGA
							CGG

1 change, U/C	1 change, U/C	2 changes, AorG/UorC A/G	1 change, A/G

asp	ser	glu	lys	lys	thr	ileu	leu
	UCU	GAA	AAA	AAA	ACU	AUU	UUA
	UCG	GAG	AAG	AAG	ACC	AUC	CUU
	UCC	CAA			ACA	AUA	CUA
GAU	UCA	CAG			ACG		UUG
GAC	AGU						CUC
	AGC						CUG

2 changes,	1 change,	1 change, A/C	1 change, A/CorU
A/CorG	GorC/A		
G/UorA			

3 Now add up the changes necessary for each sequence:

(a) $1+1+1+2$ $= 5$

(b) $1+2+1+2+2+1+1$ $=10$

(c) 0

(d) $1+1+2$ $= 4$

(e) $1+1+2+1+2+1$ $= 8$

(f) $1+2$ $= 3$

(g) $2+1+1+2+1+1+1+1+1=11$

4 Identity of organisms from which cytochrome was obtained:

(a) tuna fish;   (b) wheat;   (c) rhesus monkey;   (d) chicken;
(e) silk moth;   (f) horse;   (g) yeast.

5 Ranking order for numbers of mutations required (in increasing numbers) and for 'evolutionary distance' from man (in increasing distance):

Rhesus monkey (c)   D — 0 mutations — a primate

Horse   (f)   G — 3 mutations — a mammal

Chicken   (d)   C — 4 mutations — a terrestrial vertebrate

Tuna fish   (a)   B — 5 mutations — a vertebrate

Silk moth   (e)   A — 8 mutations — an invertebrate animal

Wheat   (b)   F — 10 mutations — a multicellular plant

Yeast   (g)   E — 11 mutations — a unicellular organism

6 The observations support conclusion (I).

Sections 21.0 to 21.3

Question 1 (*Objective 1*)

Mark the following statements as *either* TRUE *or* FALSE:

(a) Most fishes live in water and breathe through gills.

(b) No aquatic vertebrate has lungs.

(c) Teleost fish are able to alter their buoyancy by means of the air-bladder, which probably evolved from a pair of lungs.

(d) The fins of fishes are very similar in structure to the limbs of amphibians and reptiles.

(e) Crocodiles pass through a tadpole stage.

(f) Birds and mammals are able to maintain constant body temperatures.

(g) Most mammals are viviparous, which means that the egg develops inside the mother and the young are born in an advanced state of development.

(h) Some birds are viviparous, but no mammals lay eggs.

(i) The habits of birds can usually be deduced by studying their beaks, feet and wings.

(j) The habits of mammals can usually be deduced by studying their limbs.

(k) Monkeys, apes and man are all primates.

(l) Reptiles flourished until mammals evolved.

(m) A part of the brain called the cerebral cortex is very well developed in man.

(n) The only modern reptiles are the snakes and lizards.

Question 2 (*Objectives 3, 4*)

Indicate which of the advantages shown in the matrix may accrue directly from the developments listed below. Insert the appropriate number/s from the matrix below each item listed:

(a) homoiothermy

1	2	3
escape from predators	increased rapidity of movement on land	increased rapidity of movement in water

(b) flight

4	5	6
increased independence from environmental change	increased protection for the young	survival in swamps

(c) viviparity

(d) airbreathing

7	8	9
easier colonization of new environments	survival on land	drain on mother's feeding resources spread over a longer period

(e) a shelled egg

Question 3 (*Objective 10*)

Indicate, by circling the appropriate letters, which of the following statements are teleological:

(a) The lungfish breathes air and can live in tropical swamps.

(b) Reptiles lay their eggs on land so that they do not have to return to the water to breed.

(c) Mammals are viviparous to protect their young.

(d) Birds incubate their eggs, which develop only when kept warm.

(e) Birds fly away to avoid predators.

Section 21.4

Question 4 (*Objectives 1, 8*)

Indicate those items in the numbered matrix which by themselves demonstrate the three evolutionary principles listed below:

A low selection pressure

1 Darwin's finches	2 shape of whale, shark and mackerel	3 hooves of cows and horses

B adaptive radiation

4 wings of birds, bats and insects	5 beaks of birds	6 feet of birds

C convergence

7 insects' mouth parts	8 Australian marsupials	9 fur and feathers

Ignore 8 if you have decided not to read Appendix 2.

Section 21.5

Question 5 (*Objective 2*)

Indicate to which of the categories listed below, the animals in the matrix belong, by inserting the appropriate number/s from the matrix in the spaces alongside each category in the list:

(a) insects

(b) molluscs

(c) teleost fish

(d) elasmobranch fish

(e) amphibians

(f) birds

(g) reptiles·

(h) mammals

1 shark	2 snake	3 oyster	4 mole
5 beetle	6 penguin	7 gibbon	8 trout
9 perch	10 skate	11 ostrich	12 turtle
13 frog	14 pterosaur	15 caterpillar	16 bat

Question 6 (*Objective 6*)

Select from the matrix below, three pairs of statements indicating co-adaptation of some flowers and some insects.

Moths have a good sense of smell 1	Grass has light pollen 2	Snapdragons have nectaries 3
Many trees have green flowers 4	Many summer flowers are red or blue 5	Bees have legs with pollen baskets 6
Bees see ultra-violet and blue light 7	Tobacco flowers smell strongly and have nectaries 8	Roses produce much pollen 9

Sections 21.6, 21.7

Question 7 (*Objectives 1, 7*)

Carbon fixation in plants is:

(a) the reverse of glycolysis in animals;

(b) the reverse of the citric acid cycle in animals;

(c) neither of the above.

Question 8 (*Objectives 1, 7*)

Which, if any, of the following substances are common to plants, animals, and bacteria?

(a) haemoglobin

(b) cytochrome c

(c) ATP

(d) chlorophyll

(e) insulin

Question 9 (*Objectives 1, 7*)

Circle the appropriate letter(s) to show which of the following biochemical mechanisms are common to *all* living organisms (other than viruses).

(a) oxidative phosphorylation

(b) protein synthesis

(c) glucose catabolism

(d) photosynthesis

(e) RNA synthesis

(f) lipid synthesis

(g) nitrate reduction

(h) active transport

(i) ATP production

(j) CO_2 production.

After reading the Unit

Question 10 (*Objective 8*)

Indicate the scientific validity of the statements in the matrix, by writing the appropriate number(s) from the matrix in the spaces against the four descriptions, (a)–(d), below.

1 The ancestor of Darwin's finches came from South America.	2 One of the earliest stages in the development of 'life' was the formation of coacervate drops.	3 Insects cannot exceed a certain size, because of the tracheal system used in breathing.
4 Man is descended from a chimpanzee.	5 *Tyrranosaurus* was a carnivorous dinosaur.	6 *Archaeopteryx* (i.e. the earliest fossil bird) had feathers and a long bony tail.
7 The first organisms to emerge from the 'primordial soup' were chemo-autotrophes.	8 Amino acids can be formed by passing an electric current through mixtures of appropriate inorganic compounds.	9 The first life appeared in the oceans.

(a) ascertained fact

(b) almost certainly true, but impossible to prove

(c) possibly true

(d) very unlikely

Question 11 (*Objectives 1, 7*)

Indicate which of the following are terminal electron receptors in plants, and which in animals.

(a) oxygen

(b) hydrogen

(c) nitrogen

(d) nitrate

(e) phosphate

(f) sulphate

1 terminal electron receptors in plants.

2 terminal electron receptors in animals.

Question 12 (*Objectives 3, 4, 5, 6*)

Insert the most appropriate word or phrase from the given list into each of the sentences below.

(a) Reptiles were able to become fully terrestrial because they can
. . . . . .

(b) Teleosts have varied swimming habits since they can . . . . . .

(c) Flying squirrels can escape predators more easily because they can . . . . . .

(d) Many flowers have characteristic scents that can . . . . . .

(e) Birds are a very successful group because they can . . . . . .

(f) Bees pollinate many flowers because they can . . . . . .

eat nuts; breathe air; run; keep warm with fur; glide; control buoyancy; alarm predators; eat insects; lay eggs on land; produce pollen; have feet adapted to many habits; suck nectar; see very clearly; attract insects; fly by night.

You should not use any phrase more than once.

Question 13 (*Objectives 8, 11, 12*)

There are about 150 species of tortoises which live their whole lives on land. Only two of these species reach giant sizes (weight over 300 kg). One of these species lives on the Galapagos Islands and the other lives on certain islands in the Indian Ocean (such as Aldabra and the Seychelles). Mark the following statements as *either* probable *or* possible *or* improbable, as explanations for these limited distributions.

(a) Tortoises can only reach giant size in a tropical environment.

(b) Tortoises can only reach giant size if they can find certain plants to eat; these plants are confined to tropical oceanic islands.

(c) Tortoises can only reach giant size in the absence of efficient mammal predators; these are not present on isolated islands.

(d) Tortoises reach giant size only on islands surrounded by water containing large numbers of fish.

(e) Tortoises reach giant size only where there are large colonies of nesting birds on whose eggs they prey (in fact, such colonies are found on these islands).

(f) Tortoises reach giant size only where the temperature never falls below 20° C.

(g) Tortoises reach giant size only where the hours of daylight never exceed 15.

(h) Tortoises reach giant size only where there are no other herbivores eating the same plants.

Self-Assessment Answers and Comments

Question 1

(a) True—see 21.3.1.

(b) False—the aquatic reptiles (e.g. turtles) and mammals (e.g. whales) breathe through lungs—see 21.3.2 and 21.4.

(c) True—see 21.3.1.

(d) False—most fishes have fins very different from the limbs of frogs and lizards; a few fishes (lungfishes) have fins with a central axis of bone but the resemblance between these and true limbs, with hands or feet, is slight—see 21.3.2.

(e) False—when a crocodile hatches from the egg, it looks like a miniature adult—see 21.3.2.

(f) True—see 21.3.3.

(g) True—see 21.3.5.

(h) Both statements in this sentence are false—all birds lay eggs and there is one group of mammals that lay eggs—see 21.3.5.

(i) True—see 21.1 and 21.3.4.

(j) True—see 21.3.5 and 21.3.6.

(k) True—see 21.3.6.

(l) False—see Table 3, 21.2.1—the 'Age of Reptiles' (described in Appendix 1) continued for more than one million years after the evolution of mammals.

(m) True—see 21.3.6.

(n) False—there are also turtles and tortoises and crocodiles—see 21.3.2.

Question 2

(a) Certainly 4; this may lead to 7, to 8 and to 2 and through 2 to 1; it could lead to 3. But if you wrote 5, 6 or 9, the connection with homoio-thermy is so indirect that you should mark yourself as wrong. Read 21.3.3 to refresh your memory.

(b) 1, 2 and 7, leading perhaps to 8. Possibly 4. Read 21.3.4.

(c) 5 and 9. Possibly 4. Read 21.3.5.

(d) 6 and 8. Read 21.3.1 and 21.3.2. Many modern fishes that live in swamps breathe air; the ability to breathe air was essential for the colonization of land. You may have added 7 and possibly 1; both these are justifiable conclusions.

(e) Certainly 5 and 8; probably 4 and 7 are also justifiable. Read 21.3.2.

Question 3

To remind yourself about teleology, turn back to Unit 18 (section 18.3.1). Sentences b, c and e are teleological—they should be rewritten:

(b) Reptiles lay their eggs on land and so do not have to return to the water to breed.

(c) Mammals are viviparous; this habit gives increased protection to the young.

(e) Birds fly away and thus avoid predators.

You may have wondered about this last statement (e)—Birds fly away when disturbed in various ways and this behaviour leads to avoidance of predators, but it is very doubtful whether the birds' flight is *on purpose to avoid* the predators.

Question 4

A   1, 8 (You will not have answered 8 unless you have read Appendix 2).

B   5, 6, 7, 8.

C   2, 3, 4, 9.

To remind yourself about adaptive radiation, read 21.0; for convergence, read 21.4; Darwin's finches are studied in 21.1.

Question 5

(a) 5, 15—see 21.5.

(b) 3—see 21.5.

(c) 8, 9—see 21.3.1.

(d) 1, 10—see 21.3.1.

(e) 13—see 21.3.2.

(f) 6, 11—see 21.3.4.

(g) 2, 12, 14—see 21.3.2.

(h) 4, 7, 16—see 21.3.5.

Question 6

1 and 8—Moths locate tobacco flowers by their scent and then suck nectar.

7 and 5—Bees can recognize flowers by their colour; some red flowers reflect ultra-violet light and so are seen by the bees.

6 and 9—Bees collect pollen from roses. (The light pollen of grass is dispersed by the wind and is not collected by insects.)

Question 7

(c)—although some of the reactions are similar to those of glycolysis—the Krebs Cycle. If you were wrong, see *The Chemistry of Life*, pp. 241–4.

Question 8

Common to all are (b) and (c).

(a) Haemoglobin is the red pigment found in the blood of many but not all animals.

(d) Chlorophyll is only found in organisms that perform photosynthesis.

(e) Insulin is a hormone found in mammals and other vertebrates.

Question 9

Common to all are b, c, e, f, h and i.

Not common are:

(a) oxidative phosphorylation—anaerobic organisms do not perform this reaction.

(d) photosynthesis—animals do not photosynthesize, for instance.

(g) nitrate reduction—is confined to a small group of bacteria.

(j) CO_2 production—generally depends on oxidation, hence not a reaction performed by anaerobes.

Question 10

(a) 6 and 8 are ascertained facts. The fossil *Archaeopteryx* is well preserved and these details have been observed. Miller's apparatus has been used to produce amino acids in the way described.

(b) 1, 3, 5 and 7 are almost certainly true. 1, 5 and 7 happened in the past and it is unlikely that there will ever be direct evidence to prove these statements; 3 is a very likely deduction from observations on living insects.

(c) 2 and 9. These are each one of several possible interpretations of the evidence at present available.

(d) 4. Man and chimpanzees are primates and both are almost certainly descended from a common ancestor. This ancestor was probably a primitive ape.

Question 11

(a) Oxygen is the only one of this list that is a terminal electron receptor in plants and in animals.

(d) and (f) Nitrate and sulphate are terminal electron receptors in certain micro-organisms.

(b) (c) and (e) are never terminal electron receptors.

To remind yourself about these, read *The Chemistry of Life*, pp. 239–40.

Question 12

The sentences should read:

(a) Reptiles were able to become fully terrestrial because they can lay eggs on land (21.3.2).

(b) Teleosts have varied swimming habits since they can control buoyancy (21.3.1).

(c) Flying squirrels can escape predators more easily because they can glide (21.4).

(d) Many flowers have characteristic scents that can attract insects (21.5.1).

(e) Birds are a very successful group because they can have feet adapted to many habits (21.3.4).

(f) Bees pollinate many flowers because they can suck nectar (21.5.1).

Question 13

(a) Probable—tortoises are poikilothermic and are inactive in cold conditions so perhaps they would not be able to eat enough to become giants elsewhere than in the tropics. It is true of reptiles in general that species that do not live in the tropics do not grow large.

(b) Possible—but most tortoises are not fussy about what plants they eat—in fact, the giant tortoises will eat many temperate zone plants.

(c) Possible—but tortoises are well protected against most predators.

(d) Improbable—there is no obvious connection between tortoises and fish.

(e) Improbable—most tortoises are herbivorous—in fact, giant tortoises also are herbivorous.

(f) Probable—this is another version of statement (a).

(g) Improbable—it is difficult to imagine a mechanism for this.

(h) Probable—tortoises move slowly and might fail to obtain enough food in competition with active herbivores such as rabbits or antelopes.

Reading List

There are two prescribed texts:

Steven Rose, *The Chemistry of Life*. You are directed to read pp. 237 to 244 while studying 21.7, and pp. 247 to 252 while studying 21.8.

A. S. Romer, *Man and the Vertebrates*, volume 1. You should read Chapter 9 (pp. 121 to 134) after finishing the Unit, and you should browse through the many photographs and diagrams in the rest of the book, especially while studying 21.2 to 21.4.

If you wish to read one of the recommended books, you will find the following chapters are related to this Unit:

N. J. Berrill, *Biology in Action*. Chapters 15, 20 to 24, 28, 34, 37 to 40.

S. D. Gerking, *Biological Systems*. Chapters 5 and 23.

P. B. Weisz *et al.*, *The Science of Biology*. Chapters 2, 11 to 16, 32.

R. Buchsbaum, *Animals without backbones* is especially recommended for this Unit. Chapters 18, 22, 23 and 24 (all in volume 2) are relevant to section 21.5. Chapters 1 to 5 (volume 1) are relevant to 21.7—you may have read them while studying Unit 18.

Legends for Filmstrips 21a and 21b

Filmstrip 21a Photographs of birds and arthropods

1 A kingfisher (*Alcedo*) perched on a branch and holding a fish in its beak. Actual size: about 19 cm in height. Note the perching toes and brilliant colours. (Bruce Coleman and R. K. Murton.)

2 A gannet (*Sula*) on the ground. Actual size: about 95 cm in height. Note the webbed feet and the predominantly white colouration. (Bruce Coleman and Harry Thomas.)

3 *Cyclops*, a 'water flea' – a small crustacean that forms part of the zooplankton of freshwaters, swimming with its 'hairy' jointed limbs, especially the long 'antennae' at the front end. The single eye is just visible as a spot between them. The two masses of dark blobs are egg sacs. Note the joints along the body and in the limbs. Actual size: up to 3.5 cm. (Bruce Coleman and Oxford Scientific Films Ltd.)

4 *Gammarus*, a 'freshwater shrimp' – a crustacean that is common in gravel or among plants in rivers and some ponds and lakes. Note the curved body (the animal often lies on one side), the joints along the body and the many pairs of jointed limbs. Actual size: up to 15 mm. (Bruce Coleman and Jane Burton.)

5 *Porcellio*, a wood louse, seen from below. This is a crustacean that lives on land, in damp places. Note the jointed body, antennae and legs, six pairs of which are used for walking.

6 A giant millipede, a herbivorous arthropod that lives in damp places. The body has many joints and many pairs of jointed legs. Through the flexibility of the joints, this millipede is able to curl up into a spiral. Note the small red mites crawling over the millipede; these are related to the spiders. (Bruce Coleman and Jane Burton.)

7 *Lithobius*, a centipede, a carnivorous arthropod that lives in damp places. Note the jointed body and pairs of jointed legs. Compare this with the millipede (6). (Bruce Coleman and S. C. Bisserot.)

8. A scorpion, a carnivorous arthropod related to the spiders. Note the jointed body and jointed limbs. The hind end of the body carries the sting and is curved upwards and forwards. There is a large pair of limbs with pincers, used to grasp prey, and four pairs of walking legs. (Bruce Coleman and Jane Burton.)

Filmstrip 21b Photographs of molluscs, insects and flowers

9 A 'sea slug'. This mollusc is not closely related to the land slugs, but resembles them in lacking the shell that is characteristic of most groups of molluscs. The slug crawls over rocks and seaweeds eating small animals related to sea anemones. Actual size: up to 9 cm long. (Bruce Coleman and Jane Burton.)

10 *Pecten*, a scallop. This mollusc is related to the oyster but is not fixed to rocks: it can swim by flapping its two shell valves. The dark spots just inside the shell valves are eyes. Actual size: up to 12.5 cm broad. (Dr. D. P. Wilson.)

11 *Sepia*, the cuttlefish. This mollusc has an internal shell. Note the large paired eyes with the tentacles in front of them and the pair of fins at the sides of the body. (Dr. D. P. Wilson.)

12 *Ochlodes venata*, the Large Skipper butterfly, taking nectar from a flower. The two pairs of wings are folded together and the long proboscis is extended from below the head. Note the jointed legs, the knobbed antennae and the large eyes. (K. G. Preston-Mafham.)

13 *Apis*, a honey bee, taking nectar from a dandelion. The two pairs of wings are folded back and the proboscis is extended under the head. Note the large eyes, jointed legs, with pollen baskets on the last pair, and jointed hind end of the body. The dandelion is actually many small flowers (florets) each with nectaries at the end of a tube. (Bee Research Association).

14. *Bombus lucorum*, a bumble bee (or humblebee), visiting a Greater Knapweed flower. Compare it with the honey bee (13). (K. G. Preston-Mafham.)

15 *Ophrys apifera*, the bee orchid. Compare the general appearance of the flowers with the bumble bee (14). (K. G. Preston-Mafham.)

16 *Lamium album*, the White Dead Nettle. This is a common wild flower, adapted for pollination by bees, which must push forward into it to be able to suck the nectar. *Lamium album* flowers from March to November. (A-Z Botanical Collection Ltd.)

S.100—SCIENCE FOUNDATION COURSE UNITS

1 Science: Its Origins, Scales and Limitations
2 Observation and Measurement

3 Mass, Length and Time
4 Forces, Fields and Energy

5 The States of Matter

6 Atoms, Elements and Isotopes: Atomic Structure
7 The Electronic Structure of Atoms

8 The Periodic Table and Chemical Bonding
9 Ions in Solution

10 Covalent Compounds

11
12 } Chemical Reactions

13 Giant Molecules

14 The Chemistry and Structure of the Cell

15
16 } Cell Dynamics and the Control of Cellular Activity

17 The Genetic Code: Growth and Replication
18 Cells and Organisms

19 Evolution by Natural Selection
20 Species and Populations

21 Unity and Diversity

22 The Earth: Its Shape, Internal Structure and Composition

23 The Earth's Magnetic Field

24 Major Features of the Earth's Surface
25 Continental Movement, Sea-floor Spreading and Plate Tectonics

26
27 } Earth History

28 The Wave Nature of Light

29 Quantum Theory
30 Quantum Physics and the Atom

31 The Nucleus of the Atom
32 Elementary Particles

33
34 } Science and Society